D0402965

The Science of
DOCTOR WHO

The Science of DOCTOR · WHO

Paul Parsons

THE JOHNS HOPKINS UNIVERSITY PRESS
Baltimore

Printed in the United States of America on acid-free paper
2 4 6 8 9 7 5 3 1

The Johns Hopkins University Press
2715 North Charles Street
Baltimore, Maryland 21218-4363
www.press.jhu.edu

Library of Congress Cataloging-in-Publication Data
Parsons, Paul, 1971–
The science of Doctor Who / Paul Parsons.
p. cm.
Includes bibliographical references and index.
ISBN-13: 978-0-8018-9560-9 (hardcover : alk. paper)
ISBN-10: 0-8018-9560-X (hardcover : alk. paper)
1. Science—Miscellanea. I. Title.
Q173.P277 2010
500—dc22 2009037381

A catalog record for this book is available from the British Library.

The Johns Hopkins University Press uses environmentally friendly book materials, including recycled text paper that is composed of at least 30 percent post-consumer waste, whenever possible. All of our book papers are acid-free, and our jackets and covers are printed on paper with recycled content.

To Gail, with love

"The localized condition of planetary atmospheric condensation caused a malfunction in the visual orientation circuits. Or to put it another way, we got lost in the fog."

The Fourth Doctor, "Horror of Fang Rock"

Contents

Preface

As *New Scientist* journalist Hazel Muir pointed out in her review of Lawrence Krauss's 1995 bestseller *The Physics of Star Trek*, nit-picking over the scientific plausibility of a fictional TV series sounds like a recipe for possibly the most irritating book ever written.

Muir's review went on to be highly favorable. And I hope you'll find that I've done similar justice to the *Doctor Who* universe here. In fact, I hope you find it to be the most accurate, most entertaining, maybe even in places most amusing, of scientific guides to the *Doctor Who* mythos that you could wish for.

That's not to say that it won't irritate you now and again. Rest assured—there are nits here, and when I find them I pick them remorselessly. But for the most part, this is not a book of nasal-toned pedantry. Rather, it's a gathering of amazing possibilities.

When I had the idea for the book in spring 2005, shortly after the show's staggeringly successful relaunch in the United Kingdom with Christopher Eccleston playing the Doctor, I had no clue quite what was going to end up filling the bulk of the pages that you now hold in front of you. I thought I knew. I thought I had a very good idea. I've been a *Doctor Who* fan since the early years of Tom Baker, a science writer and journalist since 1996, and a keen science student and postgrad researcher for almost a decade before that. So it shouldn't have been a difficult knot to tie—make a list of the scientific themes in the series and then link them up with what's happening in laboratories around the world. Indeed, I soon had the back of a sizable envelope chock-full with some fascinating, if rather obvious ideas: time travel, cybernetics, energy weapons, regeneration, teleportation, spaceflight, and so on.

It was only when I picked up the phone and started talking to researchers that I began to realize just how much is going on in labs that I really hadn't got a clue about—and how close these topics were to the speculations of the *Doctor Who* writers. For example, in Chapter 5 you can read about the sonic screwdrivers that are already a part of modern manufacturing technology. Then there's the scientific reason that the Doctor is so nice to everyone (see Chapter 1). In Chapter 10, you've got the deflector shield for tanks that can stop a projectile just like the Daleks do in the new series. There's the real K-9—the wheeled robot assistant for space travelers that NASA is developing (Chapter 22). Or the project to record and store human memories for posterity, just like the Time Lord Matrix on Gallifrey (Chapter 29). These are to name but a few.

As Fred Taylor, planetary scientist at Oxford University, pointed out, after one of several discussions we had about the alien worlds featured in the show: "One of the charms of *Doctor Who* is that the ideas are just beyond what seems reasonable with foreseeable science, but not so far that they would never be possible. They are so often cleverly related to current thinking."

Many of the scientists whom I contacted for advice turned out to be fans of the show, as well as experts in their respective professional fields. They were able to relate their remarks, which you'll find dotted throughout the text, to particular episodes and anecdotes with little prompting from me.

Does a science fiction TV show really *need* a guide to the science behind it? No, of course it doesn't—but then we don't really *need* SF TV shows either. We watch them because they're fun, because they fire up the imagination, and because they offer something by way of an escape from life's routine. I hope you can enjoy this book for some of the same reasons.

So, make yourself comfy on the sofa—or behind it, as you like—break out the jelly babies and unplug the clock. There's someone I'd like you to meet.

NOTE TO THE U.S. EDITION: The main text of this book was written in the summer of 2005, following the screening on BBC Television of the first relaunched series of *Doctor Who,* starring Christopher Eccleston as the Ninth Doctor. I then updated the book for the UK paperback edition, looking at the science behind some of the topics raised during David Tennant's first season as the Doctor, initially broadcast on BBC TV during the spring of 2006. I have updated it again, just before the regeneration of the Doctor into his eleventh form in the winter of 2009 for this edition. Who knows, with the constant changes in science and the popularity of the television show, I may be updating it again in the near future!

Acknowledgments

Besides eleven very special Doctors, without whom there would have been nothing to write about, I would also like to extend profuse thanks to the following doctors, professors, and general good eggs for giving generously of their time and for being patient with my repeated botherings during the production of this book:

Douglas E. Adams, Ann-Christine Albertsson, Susan Aldridge, Jim Al-Khalili, Peter Atkins, John Barrow, Horace Barlow, Peter Barlow, Frank Barnaby, Martin Barstow, Françoise Baylis, Jim Bell, Halla Beloff, Mike Benton, Norman Billingham, Pete Bitar, Piers Bizony, Alfred Blumstein, Petra Boynton, André Brack, Sam Braunstein, John Brookfield, Clifton Bryant, Graham Cairns-Smith, Robert Caldwell, Robin Canup, Angela Carter, Roz G. Carter, Paolo Ciarcelluti, Stuart Clark, Bryan Clarke, Rick Claus, Frank Close, Bill Clyne, Jack Cohen, Terry Deacon, Ronny Decorte, David Deutsch, John Eades, Ulrich Eckern, Alex Ellery, Dylan Evans, John Fabre, Jim Feast, John Finney, Brian Forde, Alan Forrester, Robert Freitas, Chris French, James Friend, Richard Friend, Henry Gee, André Geim, Matthew Genge, Gerry Gilmore, Fred Glasser, Norman Glendenning, Steve Grand, Fred Haas, Mark Hadley, Wendy Hall, Craig Hoyle, Randy Hulet, Richard Jozsa, Masashi Kawasaki, Bernard Kay, Cecil Kidd, Andrew King, Tom Kirkwood, Serguei Krasnikov, Joe Lamb, Gareth Leng, Bernard Levy, Andrew Liddle, Bruce Lynn, Amanda Lynnes, Malcolm MacCallum, Norman Maclean, Ron Mallet, Ineke Malsch, Robert Matthews, Colin McInnes, Ralph Merkle, Stephen Minger, Denis Murphy, Bill Napier, Denis Noble, Igor Novikov, Kieron O'Hara, Ken Olum, Tsvi Piran, Nick Pope, Alan Rayner, Peter Robbins, Tony Ryan, Klaus-Peter Schröder, Noel Sharkey, Seth Shostak, Karl Shuker, Sharon Smart, Graham Southorn, David Stevenson, Tim Stevenson, Ian Stewart, Marshall Stoneham, Julie Suhr, Uwe Tauber, Fred Taylor, Max Tegmark, Graham Thompson, Diego Torres, Jim Tucker, Margaret Turnbull, Chris Van Den Broeck, Cees Vermeer, Ray Volkas, Mark Warner, Michael Wartell, Kevin Warwick, Irv Weissman, Christopher Wills, Richard Wiseman, William Wootters, Klaus-Peter Zauner, and Giancarlo Zema. Without their help, none of this would have been possible.

I'd also like to thank my family and friends, whom I have woefully neglected over the time that this book was being written. That includes everyone on *BBC Focus* and *BBC Sky at Night* magazines, who were forced to endure three months with a zombie in the office.

Special thanks to the Johns Hopkins University Press, for taking this project on, and to agent extraordinaire Peter Tallack at the Science Factory, for his insights and encouragement—and his infinite knowledge of the business.

Finally, I would like to take this opportunity to acknowledge the continued support of Red Bull energy drinks, McVities biscuits, and the rousing reggae sounds of Apache Indian, all of which kept my eyes open and my fingers moving during the small hours on umpteen occasions.

The Eleven Doctors

First Doctor: **William Hartnell,** 1963–66

Second Doctor: **Patrick Troughton,** 1966–69

Third Doctor: **Jon Pertwee,** 1970–74

Fourth Doctor: **Tom Baker,** 1974–81

Fifth Doctor: **Peter Davison,** 1981–84

Sixth Doctor: **Colin Baker,** 1984–86

Seventh Doctor: **Sylvester McCoy,** 1987–89, 1996

Eighth Doctor: **Paul McGann,** 1996

Ninth Doctor: **Christopher Eccleston,** 2005

Tenth Doctor: **David Tennant,** 2005–9

Eleventh Doctor: **Matt Smith,** 2009–

• Part One •

DOCTOR IN THE TARDIS

1

Who Is the Doctor?

"I've already told you. I am known as the Doctor. I'm also a Time Lord from the planet Gallifrey in the constellation of Kasterborous."
"You're bonkers."
"That's debatable."

—The Sixth Doctor and Russell, "Attack of the Cybermen"

So what exactly is it you're a doctor of, Doctor . . . ? The truth is we never really find out. Nor do we know the Doctor's real name (he variously uses "John Smith," "Dr. W," and "Doctor von Wer"—*wer* is German for "who"); or whether he has a family (though he did refer to his first assistant, Susan Foreman, as his granddaughter); or the real reason that he's wandering the fourth dimension in a time machine that has clearly passed its sell-by date (at different points in the show we're told that he's in exile from his home world, that he left to explore, and that he ran away because he was bored).

We do know that he's a 900-year-old alien from the planet Gallifrey. He's a Time Lord—one of a race that polices the galaxy, clamping down on unlicensed time travel. He's fond of planet Earth, has a soft spot for humans, and is rather keen on cricket.

The Doctor studied at Gallifrey's Prydonian Academy, where his major was thermodynamics. However, as his assistant Romana points out in "The Ribos Operation," he only just scraped through his final exams—and that was on the second attempt.

We also know that he's rather well connected. He has met Isaac Newton—we're told in "The Pirate Planet" that he dropped an apple on the great man's head and then proceeded to explain gravity to him over dinner. He helped Shakespeare write the first draft of *Hamlet* after the Bard had sprained his wrist writing sonnets. And he's met and befriended the likes of H. G. Wells and Charles Dickens.

Although initially portrayed as a tetchy old man, the Doctor soon emerges as a vastly intellectual force for good in the Universe. He fights evil and cruelty wherever he finds them, even if that means breaking the established rules or placing his own life in grave danger. Not that his life is ever in *that* much danger—he has the ability to cheat death by regenerating into a new form, with his character and outlook sometimes changing as wildly as his appearance from one incarnation to the next.

The Doctor is an enigma, a paradoxical personality. He battles against evil and yet refuses to bear arms. He shows deep compassion, yet on occasions can be utterly ruthless. He preaches responsibility and yet time and again leads his human companions into the line of fire—and occasionally to their deaths.

So can science really cast any light on what it is that makes the Doctor tick? More to the point, can it explain why he has two of them?

One Mind, Two Hearts

At times, the human body can be something of a puzzle. For example, why is it that we have two of some organs but only one of others, such as the heart? We have two kidneys. Neither kidney has a distinct function, beyond acting as a backup should the other fail. And the heart and the kidneys are both vital—the kidneys remove waste fluid (urine) from the blood, while the heart pumps oxygen-rich blood around the body to keep it alive. So if nature's gone to the trouble of giving us two kidneys, then why not two hearts as well?

The Doctor has two hearts (as do all Time Lords). As an unnamed character remarked to him in "Terror of the Autons": "We've always felt that your hearts are in the right places." So how does the Doctor's extra heart work? Is it just a spare? Or does it serve some deeper purpose? And could such a secondary cardiovascular system be of any benefit to humans?

In some ways, humans already have two hearts. Or rather, we have one heart that's made up of two pumps, each of which moves blood around one of two distinct circulation systems in the body. The first is the pulmonary system, a network of blood vessels that runs through the lungs. It's served by the heart's right atrium and right ventricle, which work together to draw in blood from the body and send it around the lungs to oxygenate it. The freshly oxygenated blood then enters the heart's second pump—the left atrium and left ventricle. These force the blood into the systemic circulation, where it's carried around the whole body.

Having these two pumps within the same organ means that their rhythms are easily synchronized. But if we had two separate hearts, their beating would need to be carefully timed so that one didn't disrupt the operation of the other.

There could be an advantage in having one heart beating while the other is relaxing. The systemic circulation system is made up of a staggering 96,000 kilometers (60,000 miles) of blood vessels—enough to circle the Earth more than twice. Forcing blood around this network with just one heart requires a huge force to be exerted with each beat, making the peak blood pressure during a beat very high. But if a second heart were to pump while the other was relaxing then the peak pressure could be much lower, making the blood flow more uniform and placing less stress on the vessels.

Alternatively, the Doctor's two hearts could allow the systemic circulation to be split into two—one system supplying the muscles, the other feeding the internal organs and the brain. When we exercise, blood is diverted from our organs to feed the oxygen-hungry muscles, compromising other body functions like digestion. A dual circulatory system could avoid this "either-or" situation and could even boost the blood supply to the Doctor's all-important brain while he's on the go.

Of course, improving the capacity of the body's oxygen transport system is of little use if the oxygen isn't there in the first place. And so the Doctor's twin hearts suggest that his lung capacity must also be enhanced somehow, if he's to oxygenate all the extra blood that his second heart is pumping around. In "Pyramids of Mars" it's revealed that the Doctor has a "respiratory bypass system." This saves him from being strangled by a mummy, and so could be interpreted as another way, other than his nose and mouth, for oxygen to enter his body.

Interestingly, when it comes to multiple hearts, it's not the Doctor who has the last word. The hagfish—an eel-like parasite found in temperate seas—has no fewer than five hearts. That's one for the brain, one for the gill pouches, one for the internal organs, and two for the tail. Fossil evidence confirms that the hagfish is the world's oldest vertebrate, surviving in the oceans now for over 300 million years—predating the dinosaurs.

But it's heart of a different sort for which the Doctor is loved best: his compassion and his unerring willingness to help others. Is there something in science that can account for altruistic behavior like this? If so, what? And why does the Doctor appear to have it in spades?

Random Acts of Kindness

Altruism—the willingness to help others to the detriment of yourself—seems to fly in the face of the modern interpretation of Darwin's theory of evolution. The theory works on the principle of "survival of the fittest," the idea that species that are well adapted to their environment are the ones best able to survive in it. Your traits—such as intelligence, physical fitness, eye color, and so on—are stored as genes, chunks of biological data written into the DNA in the nucleus of every cell in your body. DNA, a long chainlike molecule, carries our genetic information from generation to generation. When you have children, you pass on some of your genes—and the corresponding traits. If those traits are beneficial then your children will live longer as a result, attract higher-status mates, and generally enjoy better lives. They are more likely to have children who will also be successful and pass those "good" genes on again. By contrast, genes that represent less-beneficial traits are less likely to be passed on.

In this scheme of thinking, our bodies are survival vehicles for our genes. Life basically amounts to a big squabble between the genes, to determine which ones survive and which die out. This idea is often termed "the selfish gene" after the book of that name by Oxford University biologist Richard Dawkins—and it really doesn't sound like it's got much to do with altruism.

But in fact being altruistic can boost the survival chances of your genes. As mathematical biologist John Haldane once put it: "Would I lay down my life to save my brother? No, but I would to save two brothers or eight cousins." His point was that you share half your genes with your brother, so ensuring that two of your brothers survive to have children confers the same survival benefit to your genes as ensuring that you yourself survive. Similarly for the one-eighth of your genes that you share with your cousin. Sacrificing yourself to save your relatives can after all bring a survival benefit to your genes.

Evolutionary biologists call this idea *inclusive fitness*. It's not the most efficient way to pass your genes on—nowhere near as efficient as having children yourself, and so passing your genes on directly—but it can have an impact.

That's all very well, and a good reason to be nice to your relatives at Christmas, but it doesn't explain the altruism we see in the Doctor. He offers his help to anyone who needs it, often people or aliens that he's never even met before—let alone those he's actually related to. So what's going on?

The Doctor's behavior is much more likely to be a case of what's called *reciprocal altruism.* This idea was put forward in 1971 by Robert Trivers, then at Harvard University. In a nutshell, Trivers's theory states that you help others who are in need in the expectation that when, one day, you are in need yourself, those you have helped will return the favor. Again, there's an obvious survival benefit to your genes, but this time it explains altruism toward those outside your family.

There's evidence for reciprocal altruism in the natural world—in particular, in ant colonies. Ants have extra stomachs that they use to store food in and from which they will feed other members of their colony who are in need. Freeloaders are quickly eradicated from the colony—any ant that isn't altruistic soon finds itself being starved by its comrades.

But hang on—this is the Doctor we're talking about here. Surely he can't be driven by the same mentality as ants? He's altruistic because he feels deep down inside that it's the right thing to do, not because he expects something in return. Doesn't he? Christopher Wills, a biologist at the University of California, San Diego, disagrees. According to Wills, the Doctor, and anyone else who feels "warm inside" when they do the right thing, is simply experiencing a psychological reward mechanism that our brains have evolved in order to make us more altruistic.

We need that reward because we're weak. Just because we know that altruism is good for us doesn't mean that we'll want to practice it. (Usually quite the reverse.) And so nature has evolved its own bonus scheme to help motivate us—making us feel good when we do good. It's exactly the same sort of feel-good mechanism that drives us to other kinds of behavior necessary to the survival of ourselves and our genes—behaviors like eating and having sex.

The psychological pathways in the brain that trigger these rewards are there because of evolution. If it's in your genes that you will feel good whenever you do something that makes you fitter, in the "survival of the fittest" sense, then you are more likely to actually be fitter, and so more likely to pass those genes on. "What we've now got in our own psyches is a tendency for us to feel good when we are altruistic," says Wills, "because it's increased our Darwinian fitness when we've been altruistic in the past."

But if altruism is such an evolutionary advantage, then how come there are so many curmudgeons around who seem to get a kick from being decidedly unaltruistic? Wills thinks that it's a matter of what works best in a given situation. Clearly there are times in which altruism is the optimal strategy and other times when it's best to be selfish. For example, he imagines a lone selfish millionaire who, through a combination of greed and ruthlessness, has done rather well for himself. Clearly, in his case selfishness has worked very well. The millionaire goes on to have lots of children, who are equally selfish. They all eventually have children and as a result selfish tendencies begin to spread through the population.

"In this case it isn't very long before selfishness becomes disadvantageous," says Wills. "A selfish approach will really only be advantageous to an individual if that individual is rare." And so as selfishness spreads, the optimum strategy shifts toward more cooperative types of behavior, such as altruism.

This looks very much like the behavior of the Doctor's archenemy, the Master. Normally an evil, selfish loner, the Master preys upon the weak and the helpless. But put him up against an equally selfish adversary and he soon changes tack. For instance, when the Master is captured by the evil Axon race in "The Claws of Axos" and forced to help them or die, he decides to team up with the Doctor instead.

So the fact that the Doctor is a nice chap boils down to being nature's way of keeping him and his genes alive. But that's not the only thing that keeps him smiling in the face of adversity. He also seems to land on his feet rather a lot. Put it this way: how many people do you know who get to go gadding around the galaxy in a time machine with an endless supply of glamorous assistants? And if that's pure luck, then what about the way he miraculously dodges Daleks, always seems to know the right people, or just happens to be in the right place at the right time when, say, an alien spacecraft chooses to slice Big Ben in half? When it comes to luck, the Doctor seems to get more than his fair share—and then some. Now scientists are discovering why.

"Do You Feel Lucky, Doctor?"

What is luck? And why do some of us (like the Doctor) seem to have bags of the stuff, while others appear woefully bereft? Studies by Richard Wiseman, of the University of Hertfordshire in England, suggest that being lucky is less about rabbits' feet and clover leaves and more to do with the psychological outlook of the individual.

Wiseman has found that people who consider themselves "lucky" are typically extroverts. They are at ease talking to strangers and so more likely to experience chance encounters that will lead to new opportunities.

Lucky people also tend to have very positive expectations that give them the confidence to achieve their goals. In an experiment at an American school, psychologists told teachers that certain children in their class were particularly gifted. There was actually nothing special about these children at all, but when the teachers treated them as if there was, their schoolwork improved and they scored higher in intelligence tests. Proof, if it was needed, that you shouldn't underestimate the power of positive thinking.

By contrast, those who rate themselves as "unlucky" are generally more introverted and have more negative life expectations. Studies show that people in this category don't just have fewer lucky breaks in their lives; they also experience a higher incidence of serious illness and are more likely to be involved in accidents.

This all means that if the genes that you inherit from your parents play a part in determining your mental outlook on life—as they probably do—then you really can be "born lucky."

There's another benefit to being an extrovert who has a lot of chance encounters with strangers: you'll soon build up a bulging contacts book. And according to Wiseman, those with a well-developed social network of friends and colleagues also experience a lot of luck in their lives. It all comes down to the so-called small world phenomenon—the idea that it's possible to reach anyone in the world via a short number of steps, first to friends, then friends of friends, and so on. It gave rise to the phrase "six degrees of separation," the idea that just six steps in your network of friends and their friends is enough to connect you to anyone else on the planet. So when you meet someone for the first time, you're not just adding their name to your contacts book, but effectively theirs plus the names of everyone they know.

To get an idea of the figures, imagine you know 200 people and each of them knows 200 people—then that's 40,000 people who might just put a juicy opportunity your way. Of course, in reality many of the 200 people will know the same people, but it gives you idea how quickly the numbers stack up.

The Doctor certainly seems to know a lot of people and never shies away from new experiences and the unknown. He also tends to put a lot of trust in his intuition. Indeed, Wiseman finds that lucky people rely on their hunches and gut feelings a lot more than their unlucky counterparts. In a survey he conducted of over 100 lucky and unlucky people, he found that nearly 90 percent of lucky people used intuition when making relationship decisions and over 80 percent used it in their career choices as well. By contrast, the number of unlucky participants relying on intuition was less—just half as many in some of the categories that Wiseman looked at.

Getting a lucky break by being a social animal sounds plausible enough. But it can't have any influence on chance events, like winning the lottery—can it? Wiseman and colleague Peter Harris, from the University of Sussex, decided to conduct a survey to find out. First they asked participants to fill out a questionnaire that would rate them as either lucky or unlucky. Then they looked at the numbers that each participant selected for the next UK National Lottery draw. The researchers found that lucky people tended to pick certain "lucky" numbers, and that these numbers were generally avoided by the unlucky people. The question was: were these lucky numbers more likely to be drawn? The short answer: no. People whom we consider to be lucky are actually no more likely to pick winning lottery numbers than anyone else.

What the lucky people did have, however, was a higher expectation of winning. And that makes a lucky person keep playing the lottery while their unlucky friend gives up, moaning that the odds are too slim. "The optimism of lucky people is boundless," says Wiseman. "Although it's delusional, it's actually good for you because it keeps you trying."

Optimism is something the Doctor certainly knows a lot about, though his take on it is typically unconventional. In "The Armageddon Factor," the Fourth Doctor lectures to his assistant Romana: "Whenever you go into a new situation, you must always believe the best until you find out exactly what the situation's all about—then believe the worst."

"Ah, but what happens if it turns out not to be the worst after all?" she replies.

"Don't be ridiculous. It always is."

2

Time and Relative Dimension in Space, or Tardis

It's the Tardis. *My* Tardis. *The best ship in the Universe.*

—The Ninth Doctor, "Boom Town"

For sale. Tardis Type 40. Blue. Eleven careful owners. Plenty of room inside.

Who in their right mind comes up with a design for a spacecraft that looks like the type of 1960s-era phone booth that was used in police emergencies, affectionately called a police box? His name was Anthony Coburn, a young BBC staff writer who spotted just such a police box while out taking a break from writing "An Unearthly Child," the first-ever episode of the show. The idea was approved, so the story goes, largely because it meant that building a prop for the Tardis would be cheap and easy. And the fact that it simply materialized and dematerialized meant there would be no elaborate (and expensive) take-off and landing sequences to orchestrate in each episode.

The writers explained the Tardis's unusual form by introducing the idea of the chameleon circuit (or "camouflage unit," as it was initially known), which Time Lords use to make their Tardises blend into the surroundings of whatever alien world or time period they're visiting. The Tardis looks like a police box because the circuit had got "stuck" during a visit to London in 1963.

We're told that the Doctor's Tardis is a Type 40 TT. Inside, it's a sprawling network of rooms and corridors. Beyond the primary control room with its iconic console and roundel paneling, there are umpteen further chambers including quarters for the Doctor and his assistants, an art gallery, a greenhouse, a sick-bay, areas with brick-walled passages, a secondary control room (with ornate wood panels, used for a time by the Fourth Doctor), and even a swimming pool (although this was jettisoned in the Seventh Doctor adventure "Paradise Towers" after it sprang a leak).

You start to wonder how many rooms it's possible to fit into a small phone box. In fact, it's the first thing most of us want to know. So: why is the Tardis bigger inside than out?

Spacious Interior

The *Doctor Who* writers say that this is down to the fact that the Tardis is "dimensionally transcendental." However, Chris Van Den Broeck of Cardiff University in Wales has other ideas.

Van Den Broeck is a physicist specializing in Einstein's general theory of relativity, or GR. We'll talk about GR in more detail later, but for the moment suffice it to say that it's our best theory of gravity. It works by describing the gravitational force as curvature, or bending of space and time—or "spacetime," as physicists like to call it. So you can imagine the gravity field of, say, our Sun creating a large bowl-shaped dent in spacetime. The planets orbiting the Sun are then rather like marbles, rolling round the inside of the bowl. Einstein's theory amounts to a set of mathematical equations linking the curvature of spacetime to the matter it contains.

In 1999, Van Den Broeck realized that by choosing the right kind of matter to do the bending, and by arranging it in just the right way, he could use GR to manipulate space and time and

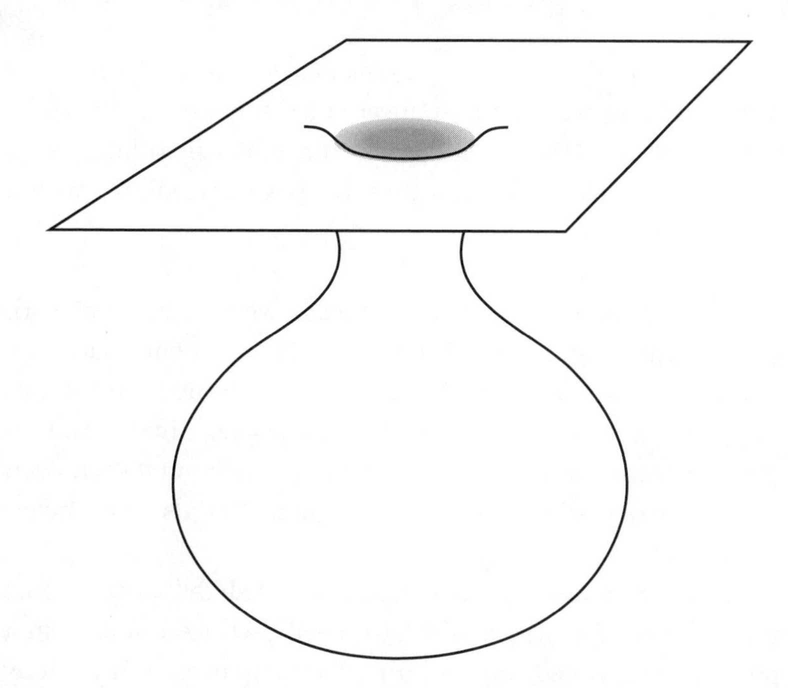

FIGURE 1. Tardis space. This two-dimensional rubber sheet has a large balloon-like bubble protruding from it. The surface area of the bubble is large, compared with the small circumference of the hole connecting it to the rest of the sheet. Move all this up a dimension and the 2D rubber sheet becomes three-dimensional space, the bubble's large area now becomes a large volume (rather like the inside of the Tardis), and the small-circumference hole becomes an entrance with a relatively small surface area (similar to the Tardis's exterior).

construct a "bubble" that's bigger inside than out—just like the Doctor's Tardis.

To get an idea of how it works, imagine a two-dimensional rubber sheet with a "bubble" attached to it (see figure 1). The bubble looks rather like a balloon, with a narrow "throat" connecting it to the rubber sheet. Now picture an ant crawling on the surface of the sheet. Where the sheet joins the throat, the ant finds a hole with a very small circumference. But if it crawls through the hole and along the throat it emerges into the bubble, a big space with a huge surface area. Essentially, the ant has encountered a large area that's enclosed by a very small circumference.

Now scale all that up a dimension, so the 2D sheet now becomes 3D space. "Replace the circumference of the throat by surface area, and the surface area of the bubble by volume," says Van Den Broeck. "What you then get is a large volume with a small surface area."

Sound familiar?

So far, so good. The problems start, however, when you work out the kind of matter that would be needed to bend spacetime in this way. Van Den Broeck found that he needed a substance that physicists call "exotic matter." This has negative pressure—pump up your car tires with it and they actually get flatter. Even more bizarrely, some varieties of exotic matter also have negative mass.

As the name suggests, exotic matter isn't all that easy to come by. But small quantities of it have been produced in labs, in a phenomenon called the Casimir effect. Discovered in 1948 by Dutch scientist Hendrik Casimir, it causes two metal plates separated by just a few millionths of a meter in a vacuum to be pulled together slightly by the negative pressure of exotic matter between them.

The Casimir effect happens because empty space isn't really empty. Instead, it's full of tiny subatomic particles, such as protons, neutrons, and electrons, popping in and out of existence over very short times. The existence of these particles, known as vacuum fluctuations, is predicted by quantum theory, the branch of physics governing the behavior of atoms and the tiny subatomic particles that atoms are built up from.

The big difference between quantum laws and the physics governing the everyday world that we are used to is that quantum laws deal only with probabilities. So rather than being able to predict, say, an atom's future state with 100 percent certainty, they would give the probabilities for every possible future state of the atom. One consequence of this is that there's a probability that pairs of particles can be created spontaneously out of empty space for short times. And these are vacuum fluctuations.

Another prediction of quantum theory is that these particles can equally well be thought of as waves. It's a mind-bending concept that physicists still don't fully understand. But there's plenty of experimental evidence for it, such as the way in which streams of subatomic particles can be diffracted (bent around corners) just like a light beam, and the way in which a light beam can impart momentum to an object that it shines on, as if solid particles were colliding with the object.

This "wave-particle duality," as it's called, is central to the Casimir effect. Outside of the metal plates, waves of all possible wavelengths can exist. But between the plates, it's a different story. Here, only certain waves can exist. The situation is rather like the waves on a piece of string stretched between two points. Because the waves must be stationary at either end, only those waves are allowed for which the length of the string is a whole-number multiple of half-wavelengths (figure 2). Similarly in the Casimir effect, the only waves that can exist between the plates are those for which the plate separation is a whole-number multiple of half-wavelengths. Switching back to particle language, this means that there are fewer particles bouncing around between the plates than there are outside of them, and so the pressure between the plates is lower than it is outside. But the space outside the plates is in a zero-pressure vacuum, and so what lies inside must have less than zero, or negative, pressure. And this is what makes the plates move together.

The amount of exotic matter produced in the Casimir effect is tiny: roughly a billion-billion-billionth of a gram. (And, yes, this mass is negative.) It's a very transient form of exotic matter—remove the plates and it goes away. But what Casimir's crucial experiment demonstrated was that exotic matter can exist within the laws of physics. All that's missing is the technology to mass-produce it.

Van Den Broeck worked out that to keep the Tardis dimensionally transcendental the Doctor needs a quantity of exotic matter roughly equivalent to the mass of a medium-sized asteroid. That's around 10^{16} kilograms (the number 1 followed by 16 zeroes) of negative-energy material.

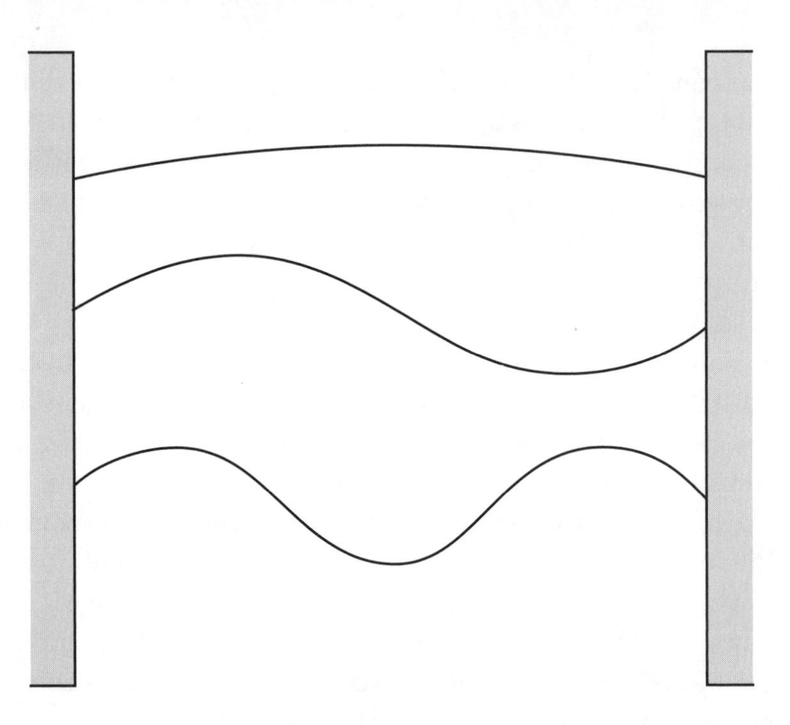

FIGURE 2. Waves on a string stretched between two fixed points—only a whole number of half-wavelengths can exist. It's the same for the "particle waves" that are allowed between two metal plates in a phenomenon of physics known as the Casimir effect.

This matter had to be arranged in a spherical shell, with a second shell slotted inside it—this one made of positive-energy matter. "The absolute values of the energies in the inner and outer layers were comparable," Van Den Broeck says. "It is conceivable that one could construct a model in which the negative energy in the outer layer exactly balances the positive energy in the inner layer."

In that case, the net total energy needed would be zero (not counting, of course, the energy requirements of whatever apparatus might be required to generate the exotic matter in the first place). Van Den Broeck stresses, however, that so far this possibility has not been investigated mathematically.

"My work was really just a toy model which captured the essentials," he adds. "It predicts that you need these negative densities but at least in principle quantum theory allows for them, so at least in principle this may be possible."

It seems, then, that we can all fit inside the Tardis. That's a good start. Now, who's got the keys?

Getting from A to B

What is it that makes the Tardis fly? Clearly, it's nothing like a conventional space rocket, with engines at the back and a cockpit where the crew sits at the front. When the Tardis "takes off" it dematerializes, fading away instead of rocketing up into the sky. That rather suggests that the Doctor's spacecraft is teleporting from one place to another—information on the state of every atom being transmitted to the destination point, where the spacecraft can then be reassembled. Amazingly, scientists have now demonstrated just this idea experimentally.

For years, it was believed that teleportation was impossible. In 1927, physicist Werner Heisenberg put forward his "uncertainty principle," one of the cornerstones of quantum theory, which says that you can never know both the position and the speed of a quantum particle, such as a proton or an electron, at the same time. The act of measuring one of these quantities disturbs the particle so much that it's impossible to know the other with any degree of accuracy. And so if you want to teleport that particle—that is, measure its state so you can then transmit that information over a distance—you're going to have a problem.

Or so it was thought. In 1993, an international team of researchers pointed out that you don't actually need to know the complete state of a subatomic particle in order to be able to teleport it. Their research hinged on an idea known as *quantum entanglement.*

Take a pair of subatomic particles and bring them together in just the right way and their quantum states (mathematical functions derived from quantum theory that describe everything about each particle, such as its position and velocity) become interlinked—physicists say the particles are "entangled." That means that if you then take the two particles far apart and measure the state of one of them, that measurement instantly determines the state of the other as well. It's rather like having two boxes—one with a red ball in, the other with a yellow ball in, but you don't know which ball is in which box. Separate the two boxes and look inside one of them—if you find the yellow ball, then you know that the other one contains the red ball.

Of course, quantum particles are more complicated than this. They have many properties and states, for example their position and speed mentioned above. But you can think of these in much the same way. So as well as red ball versus yellow ball, there will also be other states that you might want to think of as orange ball versus brown ball, others that you might call green versus purple, pink versus blue, and so on. The key property is that, whatever the state of one quantum particle, its entangled partner will always be opposite.

The researchers realized that this "oppositeness" property offers a cunning way around the Heisenberg uncertainty principle. Although the principle may stop you from knowing a particular piece of information about a quantum particle, it doesn't stop you from transmitting that particular piece of information without knowing it. (Think of the information as being in an envelope—you don't have to look inside the envelope in order to put it in the mail.) The researchers reasoned that the oppositeness must also apply even to unmeasured properties, in which case it would be possible to use it to convey information that had never been measured.

So say you want to teleport a particle (let's call it "A") from Earth to Gallifrey—here's what you do. First you join A up with an already entangled particle "B." Now you need to measure the relation between A and B. The uncertainty principle prevents you from knowing the entire quantum state of A and B, but what it does give away turns out to be just enough so that when you radio this information across to B's entangled partner "C" on Gallifrey, the receiver there can use it to extract an exact copy

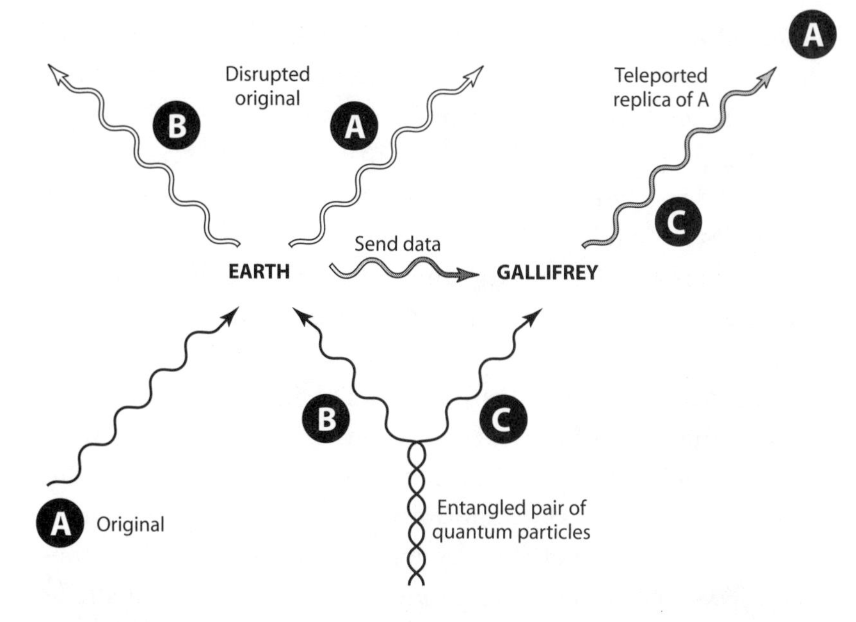

FIGURE 3. How to teleport a particle A from Earth to the Doctor's home planet of Gallifrey. A pair of entangled quantum particles B and C are separated—B is taken to Earth and C to the destination point on Gallifrey. A is then combined with B, their joint state is measured and this information is radioed across to Gallifrey. The result of this measurement tells the receiver on Gallifrey what he has to do to B's partner C in order to extract an exact copy of A. A and B, meanwhile, are completely destroyed in the process.

of particle A from C. "For example, if the measurement showed that A and B are "deeply opposite," then since particles B and C were already themselves "deeply opposite" (because that's the way they were entangled) then you know straight away the state of particle C is now the same as the original state of particle A," says William Wootters of Williams College, Massachusetts, one of the research team members. Meanwhile, the original A is destroyed in the process (see figure 3).

Another team of researchers led by Anton Zeilinger at the University of Innsbruck, Austria, proved this concept in 1997 when they used it to teleport photons—particles of light—two meters across their lab. Since then, other groups have succeeded in teleporting solid particles such as atoms. (I say "teleporting solid particles," although in fact what's being teleported is the information needed to turn the entangled partner at the destination into a copy of the original particle.) And in February 2002 two researchers from the universities of Oxford and Calcutta proposed a technique that they say could be used to entangle any kind of particle—allowing, in theory, even large molecules to be teleported.

So if you can teleport large molecules across a lab, then how soon will it be before we can send the Doctor and his Tardis to Gallifrey and back?

Don't get your hopes up just yet. The first major technical hitch is the amount of data that need to be transmitted. Let's just try to teleport a single human-size being—never mind the Tardis in all its sprawling enormity. There are some 10^{28} atoms in the human body. Let's be very generous and say that the information we need about each atom can be specified by a single bit of information (a bit is a binary digit—a "1" or a "0"). So we have 10^{28} bits of information to transmit (in reality it would be a larger figure than this). Now compare that to the fastest rate that we can relay data back from space-probes. One of the fastest deep-space data links operates between Earth and NASA's Mars Reconnaissance Orbiter spacecraft. Since November 2006, it has been beaming pictures and scientific data back from the Red Planet at a rate of 5.6 million bits per second. That's comparable to a home broadband Internet connection. But even at this speed it would still take some 10^{13} years—or about 4,000 times the age of the Universe—to download yourself. The fastest Internet backbones in use today can improve this by a factor of 100,000, but it would still take around 100 million years—quite a wait. That's the first problem.

Another is that you can't simply go where you please. Whereas the Doctor sets his controls for whatever strange planet or time period takes his fancy, all the experiments carried out so far require a receiver to be placed in advance at the destination point. So unless the Time Lords have planted receivers on every planet in the Universe, right from the beginning of time, the Doctor's not going anywhere.

And if that's not bad enough, then there's the whole philosophical question about what happens to your "soul" when you teleport. Is the soul (your sense of consciousness and identity) simply defined by the quantum state of all the atoms in your brain? Or is there more to it than that? Nobody really knows. If there's more to the soul than just physics, then teleportation is sunk yet again. Beaming inanimate objects around will be fine, but try to send a person and what you'll get out the other end will either be a soulless automaton or someone with a completely different personality from the one who went in. Then again, some physicists argue that because *only* information is being sent, then only the soul is being teleported—not the body.

The jury is still out on this one. But for the moment there are so many uncertainties that the chances of us building a real teleporting Tardis don't look good. That said, things change. No one guessed, back when the first computers were being built in the 1940s, that just 50 years later there would be this thing called the Internet—never mind the fact that we'd all be using it to do the shopping, steal music, and chat up potential partners. Maybe in another 50 or 100 years' time, new breakthroughs in teleportation technology will surprise us again.

Now You See Me . . .

Assuming that one day teleportation across the galaxy—or even through time (see Chapter 3)—does become possible, then there's another piece of Tardis technology that the Doctor may find handy: the chameleon circuit. The idea is that when you travel to ancient Egypt, it blends you in to your new surroundings by disguising the Tardis as something suitably ancient and Egyptian—say, a pillar or a sarcophagus. Otherwise it might stick out like, well, a blue police box.

The chameleon circuit is subtly different to the "cloaking devices" used, for example, by the Daleks in the Ninth Doctor episode "Bad Wolf." All a cloaking device needs to do is project a false visual image over the user—usually an image of whatever's behind them—so they appear invisible. This has been done by a team from the University of Tokyo. They've developed an invisibility cloak consisting of a silvered garment on which high-quality images from a system of cameras and projectors can be displayed. Videos of the cloak in action can be viewed on the Internet, and they are staggering (for some examples, see http://bit.ly/1zo2kb).

Although the Tokyo set-up currently requires bulky cameras and projectors, external to the cloak, innovations in camera construction (how small is the digital camera inside your mobile phone?) and "electronic paper" (technology to build display screens with the thickness and flexibility of a sheet of paper, such as those under development at the Media Lab at the Massachusetts Institute of Technology) mean that invisibility could be a practical reality before very long.

But it takes much more than a cloaking device to explain how the chameleon circuit works. This isn't just smoke and mirrors—it's more than just an image on a silvery cloak. Put your hand on the outside of the Tardis and you can feel the paneling, the windows, the cold metal of the door handle.

How do you change not just the color of materials but also their shape and texture? Tony Ryan, a polymer chemist at the University of Sheffield in England, says this could be done using a phenomenon called *electrostriction,* whereby plastics and other non-conductors can change shape under the application of an electric field—sometimes considerably. It happens when molecules in the material align themselves according to the direction of the field. Ryan's lab has developed electrostrictive plastics that can expand and contract by as much as a factor of five.

Ryan envisages the mouth of the Van Den Broeck spacetime bubble sitting inside a kind of "smart tent" made of this material. "When the Tardis lands somewhere, whoever's inside would be able to Google for an appropriate building or object, and then assemble the smart tent into that shape," says Ryan. This would presumably be done using a computer to configure an electric field in just the right way to deform the tent into a pillar, grandfather clock, or police box, as required. Exterior color could then be added by the same scheme used for the Tokyo team's invisibility cloak—although the image would be delivered via some sort of electronic paper covering the outside of the tent, rather than through external projectors. You can almost do all this using today's technology, says Ryan, never mind having to rely on devices from centuries in the future.

Having the best camouflage system in the Universe, however, won't be much use if you're only going to be forced out of hiding at lunchtime in search of the nearest McDonald's. Just as well, then, that the Tardis isn't just a spacecraft and a time machine—it's a self-service café as well.

Taste Exploration

In the William Hartnell serial "The Daleks," we not only meet the Doctor's ultimate enemy for the first time but also get to see an intriguing piece of Tardis gadgetry in action. The "food machine" takes component parts of different foods to whip up any snack or meal that a hungry time traveler might desire (albeit in the form of a somewhat unappetizing white cube). The Doctor explains that flavors are like color that the machine mixes and blends.

NASA's Institute for Advanced Concepts is investigating something very similar. They are looking for a way to produce a tasty, satisfying variety of meals that will help keep astronauts' spirits up on long-haul missions to Mars and maybe beyond—and to do so conveniently, using ingredients that could be anything up to five years old.

To help them build the device they've recruited chemist and food specialist Hervé This of the Collège de France in Paris. He's been working on a symbolic language that can be used to express an entire recipe as a single neat formula—a kind of food mathematics.

The symbols of This's language correspond to raw ingredients and the culinary steps to be performed upon them—such as whisk, bake, chill, grind, and so on. Eric Bonabeau of research company Icosystem in Cambridge, Massachusetts, is working with This to develop a suitable range of raw ingredients that capture a wide gamut of flavors and textures and yet can survive the rigors of a real space odyssey.

By programming the formulae into a production line of robot food processors, stoves, refrigerators, and pots and pans, it should in theory be possible to cook up almost anything a hungry astronaut might fancy after a hard day on the high frontier. It could even rustle up a few "chef's specials"—one idea is to let the food formulae evolve themselves inside a computer to create new recipes for mathematically perfect super grub. And not a white cube in sight.

So it looks like this is one area where scientists have actually put one over on the *Doctor Who* writers. Café Tardis, make way for haute cuisine on Mars.

The food machine has appeared only once or twice on *Doctor Who*. But something you never get to see at all is, ahem, the other end of the process. Where exactly are the toilets in the Tardis? And just how do they work?

Using a toilet during real space missions today is a sticky subject. After all, there's no gravity to usher the wastes in a downward direction. Instead, the toilets used by NASA aboard the space shuttle employ a vacuum cleaner–like device to suck away urine and feces.

A combination of footstraps and leg bars keep the user clamped to the bowl for the extraction of solid waste. Meanwhile, urine is collected using a unisex hose-like contraption with a funnel on the end—either from a sitting or standing position.

Then where does it all go? On the shuttle, solid waste is compressed and stored, to be removed upon landing, and liquid is vented into space. On the International Space Station, where water is precious, an elaborate filtering and recycling system processes all liquid waste and returns it to the drinking water supply.

Of course you can't open a window in space, so air from bathroom areas is extracted, filtered for both odor and bacteria and then returned to the crew cabin.

True, many of these difficulties don't apply in the roundeled cubicles of the Tardis. Here there is Earth-like artificial gravity at all times (see Chapter 26 for a discussion of how this might work), which makes defecatory procedures infinitely easier. But even so, where do all the waste products go? Does the Doctor dump them into space, so to speak? That doesn't sound very environmentally friendly, and anyway how does he get them outside Van Den Broeck's bubble? Maybe he saves them all up for his next fly-by of the Dalek home world, Skaro? Or perhaps this is the real source of those curious prepacked cubes that emerge from the Tardis food machine (after all, if NASA recycles liquid waste, then why not solids too)?

Whatever the writers decide, let's just hope there's plenty of reading material in there. There's a long journey ahead—into the past.

3

Into the Vortex

"Did I mention it travels in time?"

—The Ninth Doctor, "Rose"

As Groucho Marx once astutely pointed out: "Time flies like an arrow; fruit flies like a banana." Nothing quite stretches the imagination or bends the mind like pondering the consequences of traveling through time.

If you owned a time machine, imagine the sights that you could witness: the birth of Christ, the death of the dinosaurs, the fall of Troy. Or the questions you could answer. Who really shot JFK? Did the Ark of the Covenant really exist? Was Shakespeare a woman? (The photo rights alone could set you up for life.) And then there are the paradoxes. What if you went back in time and prevented your own birth? Or what if you kept going back to the same point in time over and over again—would you create an army of clones of yourself? And in that case which one of them would be the real you?

It's fair to say that *Doctor Who*'s enormous success is due in part to the Tardis's ability to flit back and forth through time. While Buck Rogers (the weapon of choice of BBC's competitor against the Doctor in the early 1980s) had to be content visiting worlds separated merely in space, the Doctor can be anywhere at any time. Over the years, he has dropped in on the Aztecs, seen the dawn of life on Earth, and skipped forward 5 billion years to watch our planet's destruction at the hands of the aging and expanding Sun.

In fact, the most popular *Doctor Who* story ever aired—the Fourth Doctor adventure "City of Death" (episode four of which pulled in a record 16.1 million viewers in October 1979)—involved no space travel whatsoever. Instead it told the story of Scaroth, the last of the Jagaroth race, who had become scattered across different epochs of Earth's history.

Yet while TV ratings are fine for fiction, when it comes to science, researchers are interested in only one thing: cold, hard evidence. So what does the evidence say about time travel? Is it really possible? And if it is, then where are all the tourists from the future?

Relative Dimensions

In 1905, a young Swiss patent clerk named Albert Einstein published something he'd been working on in his spare time: the special theory of relativity. It was a new way of looking at the dynamics of moving bodies, which Einstein had formulated in response to a number of experimental results suggesting that the speed of a light beam, as measured by an observer, is independent of the speed that the observer is moving at. If you're in a car doing 100 kilometers per hour and you pull alongside another car doing the same speed, then the relative speed between the two cars is zero. But special relativity asserts that this commonsense view doesn't apply to light and that light always travels at the same speed relative to the observer (300,000 km per second) regardless of how fast he or she moves. One of the theory's main predictions was an effect called "time dilation," which predicted that moving clocks tick slower than stationary ones. It happens in order to keep the speed of light constant as the observer travels faster and faster. Stretching out the interval between successive ticks of a clock in the observer's frame of reference means that a light beam can travel further in the space of each tick—so it appears to move faster. So the faster you travel, the faster the light beam travels—so that its speed relative to you remains constant. At everyday speeds, time dilation is negligible—so don't expect your watch to slow down when you get on the bus. But accelerate up to the speed of light and odd things start to happen. For instance, if you were on a spaceship moving at nine-tenths the speed of light (0.9c, as physicists would say), then a "minute" on your watch would last twice as long as a minute on the watch of someone who wasn't moving.

Let's say that you rocketed away from Earth at 0.9c and traveled for six months before turning round and coming back. Your wristwatch and your wall clock all tell you that a year has passed. And your biological clock—that is, your body—has aged by a year. But on your return to Earth you find that in the time you were away, two years have elapsed there. That's a year more than you've experienced—you have jumped a year into the future. The faster you travel, the more pronounced the effect becomes—allowing, in principle, rapid journeys into the far future.

Time dilation is verified on a daily basis in the world's subatomic particle accelerators. In these giant underground tunnel rings, some of which are many kilometers across, subatomic particles are whirled round at close to lightspeed and then smashed into one another. It's not just the physicists' answer to WrestleMania—the idea is to study the fragments from the collisions to learn more about the composition and behavior of the particles involved. The researchers find that some of the fragment particles produced in the collisions are unstable and decay into other particles over well-known timescales. But these decay times are longer than they would be if the particles were stationary. Because of their high speeds, the particles' decay times have been stretched out—exactly in accordance with the law of time dilation, as predicted by Einstein.

The only trouble with traveling into the future this way is that it's a one-way trip. Time dilation can't help you travel into the past, and so there's no way to get back to your own time.

The first steps to lift that restriction were made in 1915 and, once again, Einstein was the one to make them. He was trying to modify special relativity (SR) to take account of gravity. At that time, the best theory of gravity was the one devised by Isaac Newton in 1687. But attempts to marry the two theories soon fell flat. The big problem was that one of the cornerstones of SR is that nothing can travel faster than light. But Newton's law has gravity taking effect on distant bodies instantaneously—that is, traveling infinitely fast.

Einstein had a radical solution. Special relativity describes space and time as "flat." But he found that by bending spacetime in just the right way he could incorporate the gravitational force (see Chapter 2). And his new theory, called general relativity (GR), could explain things that Newton's couldn't.

One such mystery solved was the orbit of the planet Mercury. In strong gravitational fields the predictions of Einstein's theory differed wildly from Newton's. Close to the Sun, where its gravitational field is strongest, the orbit of Mercury does a strange thing. The overall elliptical shape of the orbit (not just the planet itself) rotates around the Sun very slowly.

The effect is called "precession" and only the formidable mathematics of general relativity theory could explain it. GR is still today our best theory for explaining why apples drop from trees.

So if you can bend space and time, then what if time could be curled back around on itself so far as to allow travel into the past? Physicists refer to such temporal loops as "closed timelike curves." And it turned out that there was nothing in general relativity to prevent them.

Wormholes to the Past

By 1916—just a year after the theory of GR was published—Austrian physicist Ludwig Flamm had already used it to build the first mathematical model of what we would today call a wormhole—a tunnel through space and time. Flamm discovered that the solution to Einstein's equations of GR describing the gravitational field around a star could be extended into the core of the star and on through into a new region of space. The mathematics describing the space around a star—or any uncharged, stationary spherical object—was derived earlier that same year by German mathematician Karl Schwarzschild. His equations gave the strength of the gravitational field at a distance r from the star. Flamm found that Schwarzschild's solution was still valid if r was replaced by $-r$, suggesting to him that there could exist objects in space that it's possible to travel into (to $r = 0$) and then out of the other side and into a new region of space where r is negative.

Einstein himself, along with his colleague Nathan Rosen, developed the idea further in the 1930s. So-called Einstein-Rosen bridges were seen as shortcuts linking vastly separated regions of space by a short tunnel (see figure 4). The term *wormhole* itself wasn't coined until the 1950s by the eminent Princeton physicist John Wheeler, who likened Einstein-Rosen bridges to the tunnels made by a worm burrowing into an apple—the straight-line distance through a tunnel from one side of the apple to the other being shorter than the distance around the apple's surface.

So wormholes could maybe give us a way to make speedy hops across the galaxy. But what's this got to do with time travel? The answer to that was provided in 1986 by a team at the California Institute of Technology headed by the renowned relativity researcher Kip Thorne. And it's beautifully simple. Imagine a wormhole, the two mouths of which are positioned side by side. Now bundle one of the mouths aboard a spacecraft and launch it off into space at 0.9c. Just as we saw earlier with our hypo-

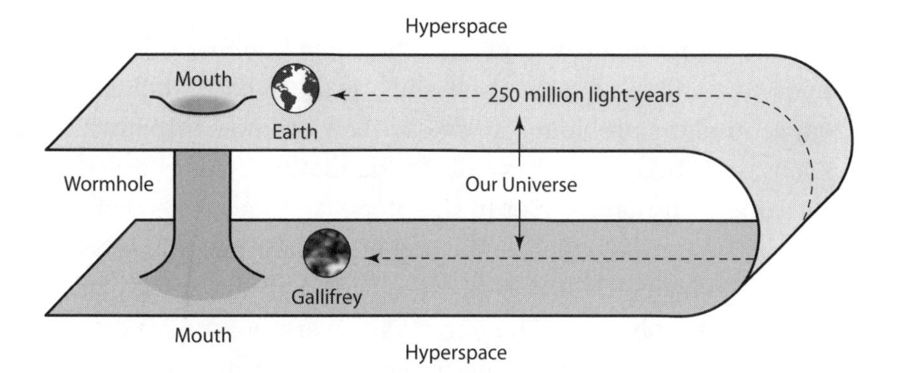

FIGURE 4. A short wormhole, just a few kilometers long, offers a handy shortcut between Earth and Gallifrey. The two worlds are separated by 250 million light-years in normal space.

thetical astronaut traveling into the future, time dilation slows down the clocks in the moving wormhole's frame of reference. And so when the two mouths are brought back together a year later, the moving one has jumped into the future by a year. But here's the good bit. Whereas before there was no way to get back to the past, bending a wormhole sets up a two-way tunnel. So if a traveler now jumps into the future mouth, he or she will emerge from the other a year earlier.

You might, quite reasonably, ask exactly how one goes about carting a wormhole mouth out into space at near lightspeed, and I'll leave the details to a suitably advanced space-faring civilization. However, broadly speaking, wormholes are areas of extreme spacetime curvature, and so—remembering that in general relativity spacetime curvature equates to gravity—they therefore have intense gravitational fields. That means that if you put a sufficiently large mass next to one—a small planet, let's say—then the wormhole will be gravitationally attracted to it. Now gently propel your planet away into space and the wormhole will follow behind. Think of the gravitational force between the wormhole mouth and the mass as a kind of space towbar, pulling the wormhole along.

The real snag is holding the wormhole mouths open long enough for people to travel through them. Studies have shown that the throat of a wormhole is rather like a rubber tube that tends to want to squeeze itself shut, sealing the tunnel. One way around the problem is to thread the wormhole throat with a substance that physicists call "exotic matter." This is exactly the same stuff that we met in Chapter 2 and that is needed to sustain the Tardis "bubble"—the region of spacetime that is bigger on the inside than it is outside. So the Tardis already has a copious supply of it. The benefit for wormholes is that the negative pressure of exotic matter generates negative gravity—or "antigravity"—to wedge the two mouths open. Current estimates suggest that sustaining a human-size wormhole requires about ten Jupiter masses of exotic matter.

Wormholes aren't the only way that theoretical physicists have come up with to travel through time. They've found Tardis-like routes to the past in all manner of cosmic phenomena, including colliding relics from the Big Bang called cosmic strings and even real-life designs for spacecraft "warp" engines (see Chapter 25). The first solution of Einstein's equations of general relativity to feature closed timelike curves actually described a variety of universe that was rotating. It was found by Hungarian-Austrian mathematician Kurt Gödel in the 1940s. Yet wormholes seem to be the closest thing in physics to the time vortex that the Tardis is so often seen hurtling through in the show.

But even with all the exotic matter in the galaxy, there's still a small technical hitch: you can only visit time periods that happen after the time machine was created. So if one was built today, you could never use it to visit the Aztec Empire, say. Think of the wormhole as a road, rather than the car that travels on it—you can't drive to places where the road doesn't go. That means that if the Tardis is essentially a wormhole generator, then it must be a very, very old one—in the Fifth Doctor story "Castrovalva" the Tardis gets caught up in Event One, the name given by the *Doctor Who* writers to the very birth of our Universe (see Chapter 30).

Wormhole-like phenomena have cropped up elsewhere in the show. In the Fourth Doctor adventure "Full Circle," the Tardis is drawn into E-Space—an alternative universe (see Chapter 32)—via a spacetime tunnel called a *charged vacuum emboitment.* Physicists have found that there is a wormhole-like solution to the mathematics describing black holes that arises when the black hole is electrically charged. An uncharged black hole has a pointlike "singularity" at its core—a region of infinite density, the gravity of which will rip apart anything that gets pulled in (see Chapter 31). But in a Reissner–Nordstrom black hole (named after the German Heinrich Reissner and the Finn Gunnar Nordstrom, who independently developed the idea in the second decade of the twentieth century), electrical charge changes the black hole's structure. A skilled spacecraft pilot would be able to steer a course around the central singularity and into a new region of spacetime entirely—an alternate universe, much like E-Space.

How the Tardis actually conjures up a wormhole each time it needs to make a journey through time isn't very clear, though efforts to build a working theory of "quantum gravity" could offer a clue. Quantum gravity is the theory that will one day replace Einstein's general relativity, combining gravity with the laws governing the tiny particles of the subatomic world. Researchers hope it will tell us what really happened during the Big Bang (see Chapter 31), when the small, dense, newborn Universe was governed by gravity and quantum theory in equal measure.

Attempts to formulate a theory of quantum gravity have been plagued by inconsistencies and impossible "divergences"—where the theory predicts that the values of real physical quantities, such as energy and momentum, soar to infinity. But promising developments are now being made in string theory (which supposes that subatomic particles are made up from vibrating one-dimensional strands of energy) and its big brother M-theory (short for Matrix theory, Mystery theory, or Muffin theory—depending on who you talk to). M-theory's main thrust is to generalize the one-dimensional objects of string theory into p-dimensional objects known, amusingly enough, as p-branes (where setting $p = 0$ gives a particle, $p = 1$ gives a string, $p = 2$ a "membrane," and so on).

If gravity really does obey the laws of quantum physics on the smallest scales, then the Tardis could be in business. In the same way that space on small scales is bubbling with subatomic particles (as we saw when discussing the Casimir effect earlier in this chapter), so spacetime itself on the smallest scales will then be a frothing mass of loops, bubbles—and wormholes. Kip Thorne at the California Institute of Technology, the researcher who first came up with the idea of using wormholes for time travel, speculates that an advanced civilization could reach down and grab one of these microscopic wormholes from the quantum realm. By threading it with exotic matter, they could then blow the wormhole up to macroscopic size, forming a stable gateway large enough to travel through.

Owning a roadworthy wormhole is one thing. Now all you need to know is how to go about driving it. And here there may be some further complications. Research published on the Internet (http://bit.ly/4zxyw6) in May 2005 by Stephen Hsu and Roman Buniy of the University of Oregon suggests that it's extremely difficult to get the destination mouth of a wormhole to go exactly where you want it to.

Wormholes come in two types—quantum and semi-classical. As the name suggests, quantum wormholes are governed by the laws of quantum physics. But these laws deal only with probabilities, and so the wormhole's behavior can't be predicted with certainty—all we can say is that there is x percent chance of it doing this or y percent chance of it doing that.

Semi-classical wormholes, however, are much more predictable, because the probabilistic laws of quantum physics play only a small part in their behavior. And that means that if you know what the wormhole's doing now, then you can say with pretty good accuracy what it's going to do in the future. That's why it's always been assumed that the semi-classical variety of wormhole is the one best suited to traveling through. You can point it at Hastings in 1066, knowing that's exactly where you'll end up.

Now, Hsu and Buniy could have put a wrench in the works. They claim to have found that semi-classical wormholes are unstable—the things literally fall apart before you can use them. This forces travelers to turn to quantum wormholes instead. And that's a nightmare, as you have no idea exactly where—or when!—they're going to drop you next.

Perhaps fans of *Doctor Who* won't find this altogether surprising, though. The Tardis is renowned for dumping the Doctor and his companions centuries and often light-years off course. During the early William Hartnell years, it was extremely unpredictable—in the adventure "The Smugglers," the First Doctor even admits: "I have no control over where I land. Neither can I choose the period in which I land."

Although the Doctor's ability to navigate the Tardis improves somewhat in his later regenerations, it remains a temperamental machine to control. In 2005's "The Unquiet Dead" the Ninth Doctor attempts to show his assistant Rose the city of Naples in 1860. Instead, they arrive ten years later—in Cardiff, Wales. Things go wrong again in "Aliens of London" when he promises to take Rose home to visit her mother and to get her there just twelve hours after they had left. She arrives twelve *months* after, to find herself the subject of a missing person's investigation.

The Blinovitch Limitation Effect

When things go badly, horribly wrong for the Doctor, as occasionally they do, why doesn't he just go back in time and put them right? Surely getting a second, third, or fourth crack of the whip is one of the perks of owning a time machine.

Admittedly, it would make for rather undramatic TV. And that's why the show's writers have come up with "the laws of time," a set of rules that curtail the Doctor's freedom to fiddle with the past. Laid down by the Time Lords themselves, the laws are:

1. Individuals are not allowed to meet themselves.
2. Individuals are not allowed to interfere with their own timelines.
3. The Blinovitch limitation effect is not allowed to happen. The Blinovitch limitation effect, or BLE, is a physical law in the *Who* universe. If there's a repeated attempt to alter history, the BLE will kick in with sometimes devastating results. This law of time essentially states that Time Lords should endeavor to stop repeated attempts to alter history and so prevent the Blinovitch limitation effect from happening.
4. No individual is allowed to travel back in time on Gallifrey, the Time Lord home world. The Time Lords consider the history of Gallifrey too crucial to the invention of time travel to risk jeopardizing.

Obviously, these are just made-up rules that have nothing to do with science. (However, they have been violated on numerous occasions, most notably "The Three Doctors," "The Five Doctors," and "The Two Doctors," in which the Doctor has teamed up with one or more of his earlier incarnations.)

The Blinovitch limitation effect, however, appears to be a law of physics in the Whoniverse. And it's one that may well be echoed in the physics of our own Universe too. Interesting as time travel is, most physicists hate it. They find the paradoxes and inconsistencies that it throws up abhorrent. So much so, when they discover closed timelike curves in their calculations, they often brand the region of spacetime containing them as "sick."

Topping the list of temporal warts they must contend with is the famous granny paradox. What if you went back in time and murdered your grandmother, thereby preventing the birth of one of your parents? You would never have been born, so you couldn't have killed your granny, so you would have been born after all, and so on. Then there's the bootstrap, or "free lunch" paradox. What if a generous time traveler took Shakespeare's complete works back in time and made a gift of them to the young Bard, who then copied them into his own hand and published them? In this case, where did the creative spark for all Shakespeare's great plays, poems, and sonnets actually come from?

Stephen Hawking, the eminent cosmologist and physicist at the University of Cambridge, finds time travel and its associated paradoxes so unpalatable that he's postulated his own version of the Blinovitch limitation effect, to "keep the world safe for historians." Hawking calls it the chronology protection conjecture. He believes that the laws of physics conspire to either destroy a time machine before it can form or to destroy anyone or anything that attempts to use one.

Hawking is yet to find a solid grounding for his conjecture within the laws of physics. He's suggested that the same vacuum fluctuations responsible for the Casimir effect could become magnified to terrific energies as they repeatedly loop through the time machine, frying anything that attempts to pass through it.

Not everyone thinks that time travel paradoxes will present a problem, though. Igor Novikov of the Theoretical Astrophysics Center, Copenhagen, has conceived what he calls self-consistency principles, which demand that there is always a non-paradoxical, self-consistent sequence of events.

To see how, imagine a simplified version of the granny paradox, in which a billiard ball goes into the mouth of a wormhole and back in time. What if the ball is aimed in just such a way so that when it emerges from the other wormhole mouth it collides with its earlier self, preventing it from entering the time machine in the first place? Novikov's self-consistency principles say that no matter how you aim the ball to begin with, there is always a self-consistent trajectory in which the emerging ball

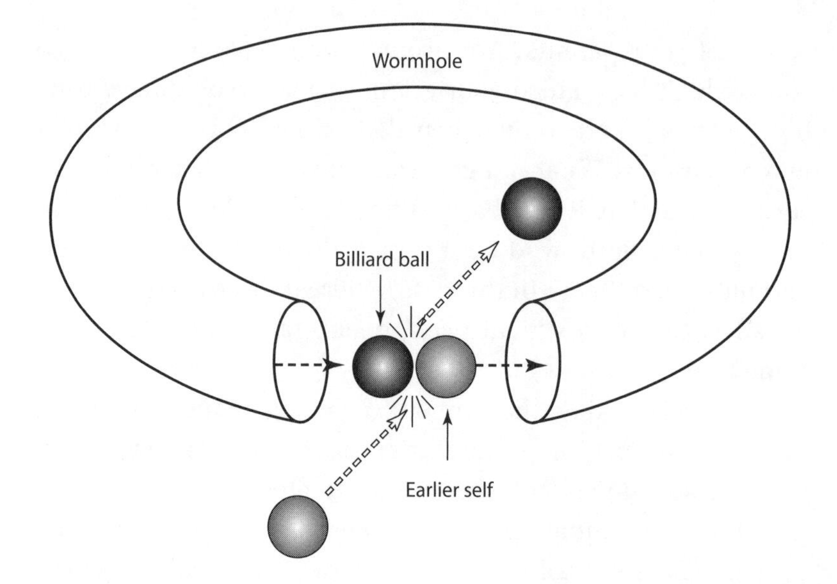

FIGURE 5. How the self-consistency principle is reflected in a billiard-ball version of the granny paradox, a construct in time travel in which a person travels back in time, kills his grandmother, and prevents his own birth. The idea is represented here in terms of billiard balls where a ball emerging from one mouth of a wormhole stops its earlier self from entering the other mouth. Self-consistency says that the ball that emerges from the wormhole is traveling at exactly the right speed and in just the right direction to knock its earlier self in.

knocks its earlier self *into* the time machine rather than away from it (see figure 5).

Novikov and his colleagues argue that self-consistency is a natural consequence of the principle of least action, one of the deepest tenets in physics. This states that for any physical system—like a ball rolling down a hill or a neutron bouncing off an atom—the "action" (which, very loosely speaking, is a measure of the system's total energy) is as small as it possibly can be. From this, physicists have been able to derive Newton's laws of motion, James Clerk Maxwell's theory of electromagnetism, and Einstein's relativity, among many others. If such a powerful principle really is trying to tell us that self-consistent, paradox-free time travel is possible, then perhaps we should take note.

Someone who does seem to have taken note is *Doctor Who*

scriptwriter Louis Marks. In his Third Doctor serial "Day of the Daleks" (which was also the first adventure to mention the Blinovitch limitation effect), an assassin is sent back from the twenty-second century to present-day Earth to kill Sir Reginald Styles—a diplomat who sets off an explosion at a peace conference, triggering a devastating future war with the Daleks. But, as the Doctor realizes, it isn't Styles who causes the explosion at all—but Shura, the assassin sent back to kill him. "You went back to change history," says the Doctor. "But you didn't change anything. You became part of it."

Even though "Day of the Daleks" first aired in January 1972, it wasn't until 1983 that Novikov first published his findings on self-consistency. Maybe that theory has now made its presence felt in a way neither he nor the *Doctor Who* writers saw coming.

4

Regeneration

"Every cell in my body's dying."
"Isn't there something you can do?"
"Yeah, I'm doing it now. See, Time Lords have this little trick, sort of a way of cheating death . . . Except it means I'm gonna change."

—The Ninth Doctor and Rose, "The Parting of the Ways"

You don't really need a Tardis to travel through time. All of us are time travelers of a sort, plodding forward at the steady pace of one year every year. We are born, we progress through childhood and adulthood, we grow old, and then we die.

Mortality and death seem to be ever-present elements of the *Doctor Who* mythos. When the Ninth Doctor, at the end of "The Doctor Dances," rejoices because "Just this once, everybody lives!" he wasn't kidding. More often, the unfolding of an adventure involves characters departing in body bags. Several times this has included the Doctor's own assistants. Adric died in the Fifth Doctor serial "Earthshock" when the Cyberman spacecraft he was aboard crashed into Earth (wiping out the dinosaurs along the way). And Sara Kingdom, a Space Security Agent who had joined the Doctor in his fight against the Daleks, was killed by "friendly fire" when the First Doctor unleashed a time destructor—a device that accelerates time—against his archfoes in "The Daleks' Master Plan."

The Fifth Doctor's companion Tegan leaves for the express reason that she can't bear all the killing any more. An entire subtext of the 2005 series looked at the Doctor's relationship with death and how he always seems to leave a wake of devastation behind him wherever he goes.

Not that any of this bothers the Doctor much. He doesn't have to face death—at least not yet. The Doctor and all Time Lords have a way of cheating the end. Called regeneration, it lets him rejuvenate every cell in his body, completely changing his appearance along the way. The process is first referred to as "renewal" when Patrick Troughton took over from William Hartnell. It wasn't called regeneration until "Planet of the Spiders," when Third Doctor Jon Pertwee handed the baton to Tom Baker.

Time Lords get 12 regenerations. So there will be a total of 13 Doctors, of which we're on number 11—Matt Smith has just taken over from Tenth Doctor actor David Tennant, who regenerated at the end of the 2009 Christmas Special, "The End of Time." But don't panic. The Doctor's evil Time Lord antithesis, the Master, was already on his thirteenth life when we first met him in "Terror of the Autons." And he has regenerated at least five times since then.

There's a good case for regeneration being the most important aspect of *Doctor Who*. How many other fictional dramas have managed to maintain the same central character so cleverly for over 40 years? The concept has also been combined with the show's other trademark element—time travel—to bring about a number of adventures in which two or more of the Doctor's incarnations have met. Regeneration has made all that possible, but could it work for real?

Ask a freshwater hydra. These cylindrical life-forms, measuring anything between 1 and 20 millimeters in length, and that you can find in any pond, are able to regenerate damage and regrow body parts. "If you were so unpleasant to a hydra as to cut it up into little pieces, you wouldn't actually be doing it a bad turn," says Tom Kirkwood, a gerontologist at Newcastle University, in England. "Each of the pieces would soon regenerate into a new organism."

The process, called *morphallaxis,* normally takes three to four days. It works using an idea in biology called differentiation. When an embryo of a life-form is first created, the cells in its body are in an "undifferentiated" state. These stem cells, as they are known, later become differentiated into types of tissue—for example, organs and skin. When a hydra regenerates, some of its cells revert into an undifferentiated stem cell state and then differentiate again into a new type of tissue. This allows damaged tissue of one type to be repaired by tissue of another and for organs to rearrange themselves and regrow. The resulting regenerated creature is normally smaller than the original, but fully functional.

The Doctor's regeneration procedure seems to vary each time he does it. Sometimes he morphs gradually from one form into the next. Other times he glows brilliantly, his new appearance emerging as the glare fades. But, if this is the same process by which hydras regenerate, then in each case the cells of the old Doctor are becoming undifferentiated and then redifferentiating into the form of the new one, healing the injuries that killed him. Presumably there must be something going on with the Doctor's genes and DNA at this stage as well, in order for his appearance to alter. Whatever this is, however, it has no analogue in the realms of earthly biology as we understand it today.

That's the theory, anyway. In practice, the situation when you're dealing with a complex human being—and probably a complex Time Lord too—is much trickier than it is for a hydra, for several reasons. Firstly, the structural differences between a tiny waterborne creature a few millimeters in size and a large mammal almost 2 meters (about 6 feet) in size are immense. Human physiology is vastly more complicated than a hydra's. Secondly, and more importantly, our cells are much more strongly differentiated than a hydra's. That means it's much harder for one kind of cell to turn into another—say, for skin cells to turn into bone cells and regrow a severed finger.

But even if there was a way around those difficulties, making it possible to regenerate a new body from old, Kirkwood still thinks there'll be one very tough problem left to solve. "The real hard part," he says, "is how you recreate the brain, and the network of neuronal connections that define memory."

Regenerating *a* brain is one thing. But that's not good enough. If you're going to regenerate, you want *the* brain—*your* brain. That means your personality, your mental abilities, and with all the connections between brain cells that constitute your own memories. Put simply, doing this is going to be darned hard. The human brain has around 10 billion data-intensive neurons—those are the cells that are important for brain function—with a staggering 10 trillion connections between them. How exactly all this information could be read, stored, and then reintroduced to a regenerated body isn't clear. (Although some steps in the preservation of memories are being made by the "Memories for Life" project—see Chapter 29.)

Perhaps this difficulty in "reprogramming" the post-regeneration brain goes some way toward explaining why the temperament and personality of one Doctor is often so different from that of the next. Perhaps it's even to blame when problems crop up—for example, when the Fifth Doctor experiences "regeneration trauma" in "Castrovalva."

Live Forever

If regenerating our bodies (or rather regenerating our bodies without losing our minds) is so hard, then perhaps we could be content with simply slowing down the aging process? In the Fourth Doctor serial "The Pirate Planet," Queen Xanxia has built a "time dam" that can slow down the passage of time for her body, effectively halting her aging process. Futurologist Ray Kurzweil thinks something very similar is possible. In fact, in an interview he gave early in 2005, he told *New Scientist* magazine in no uncertain terms: "I'm not planning on dying."

Our bodies age because every minute of every day a war is being fought in our cells, between the agents of damage and repair. Leading these agents of damage are the so-called free radicals, atoms or molecules that are particularly reactive with other chemicals. They play an important role in various life processes, such as killing bacteria inside cells. However, because they are so reactive they cause damage to the DNA in the nuclei of healthy cells as well. Because of this, the body has evolved defense mechanisms in the form of enzymes (proteins that accelerate chemical reactions) that limit the damage caused by free radicals and do their best to repair it.

"Every day of your life in each cell of your body, the DNA is zapped 10,000 times or more by oxygen free radicals," says Kirkwood. "The great majority of those zaps are repaired—say, 9,997 of them are put right. But those few that aren't fixed accumulate, and it's those that cause aging."

This all seems fairly inexorable. Does Kurzweil really think he can stop the rot? He's certainly no fruitcake. This is the man who predicted the Internet revolution back in the 1980s. He invented the flatbed scanner. He developed the first text-to-speech synthesizer. And in 1999 he was awarded the National Medal of Technology—America's highest technological honor. So when Kurzweil thinks he's onto something, you tend to want to sit up and listen.

He's been led to his enthusiasm for immortality by the rate at which biotechnology is currently developing. It's now exponential, typically doubling in power every year or so. Kurzweil believes this rapid growth can be exploited by anyone who wants to live forever. He calls his approach "a bridge to a bridge." The basic idea is that you use the technology that's available today to prolong your aging process for as long as possible. All you need to do is hold back the years until some other new technology becomes available to see you to the next bridge. And so on.

As Kurzweil told *Wired News* in February 2005, every day he takes 250 dietary supplements, eight to ten glasses of alkaline water, and ten cups of green tea. And he avoids every form of vice, including coffee. This is his "bridge one." He believes it will minimize the effect of free radicals and keep his middle-aged body in the condition of a 40-year-old long enough for the next big advance to arrive.

What will that be? Bridge two, he says, will use various medical technologies to assess the likelihood of his developing any serious illnesses, such as cancer, and then select the most appropriate therapy. These technologies will include tests to spot which diseases he is most genetically prone to—tests that are already becoming available.

Bridge three involves the revolution in nanotechnology that many techno pundits (including Kurzweil) predict is due in about 20 years. Kurzweil believes that when this happens he will be able to inject a swarm of nanorobots into his bloodstream (rather like the nanogenes featured in the Ninth Doctor episodes "The Empty Child" and "The Doctor Dances"—see Chapter 22). Each just a few thousandths of a millimeter across, these tiny devices will take the place of his digestive system, extracting exactly the right balance of nutrients from his food and delivering them precisely where they're needed around his body.

Who knows what's in store as regards bridge four? But after all that, let's just hope Kurzweil remembers to look both ways when crossing a bus lane.

Not everyone shares Kurzweil's optimism, however. "You'll find that there are plenty of people who will say that in 25 years' time we will be able to make people live forever," says Kirkwood. "I disagree profoundly with that view."

As well as expressing doubts over the plausibility of such claims, Kirkwood questions the motives of those who seek immortality. He points out that over the last 200 years life-expectancy has steadily increased by about 2 years per decade and argues that the pressing issue now is not how to increase it further but how to improve the quality of those extra years that we've already gained. "We should first be trying to delay the onset of the more disabling or unwelcome side effects of the aging process," he says. "When we've shown we can make real progress in this direction, then might be the time to start thinking about slipping in some more good-quality years as well."

The quest to live forever all sounds rather reminiscent of the sorry tale of Gallifrey's Lord President Borusa, in "The Five Doctors." Borusa sought immortality, in the form of unlimited regenerations, from the ghost of Rassilon—the founder of Gallifreyan society, who first developed Tardis technology and led his people to become lords over time. Rassilon granted Borusa's wish—freezing him into the stonework on the side of his tomb.

5

One Giant Leap for DIY

"Who looks at a screwdriver and thinks, 'oo-hoo, this could be a little more sonic*?'"*

—Captain Jack Harkness, "The Doctor Dances"

The sonic screwdriver is the Doctor's answer to a Swiss Army knife. First appearing in the Second Doctor serial "Fury from the Deep" in 1968, when it was used simply to undo the screws on an oil pipe, it's been put to some rather more complex uses since—from opening locks to welding metal to reversing the effects of teleportation—and has got the Doctor out of more than a few tight spots.

This small silver cylinder, just slightly larger than a fountain pen, has become an icon of the show. But how does it work? Is it possible to turn screws from a distance using just a beam of sound? And would that really be any help at all when it comes to doing up your bathroom?

According to Douglas Adams of Purdue University, Indiana, who is an expert in noise and vibration engineering (but no relation to the late, great *Doctor Who* script editor of the same name), the sonic screwdriver could work using what's called the principle of structure-acoustic linear ultrasonics.

It sounds like a piece of technobabble that even Jon Pertwee's

Third Doctor would be proud of, but the basic idea is remarkably simple. "You produce a focused column of oscillating air particles (sound waves) that's directed toward an object, say a screw," says Adams. "These oscillating air particles set up high-frequency vibrations in the screw, causing it to rattle along in the direction of the threads to either tighten or loosen." Which way the screw actually turns is determined just by rotating the sound beam either clockwise or counterclockwise accordingly.

This kind of sonic manipulation is already finding practical applications. For instance, it's being investigated as a replacement for the motor bearings in computer hard drives. The disks in the drive are levitated and rotated entirely using sound. Or rather, ultrasound—acoustic vibrations with a frequency above 20 kilohertz, the upper limit that the human ear can perceive. And there's a good reason for this. Delivering enough oomph to spin the disks requires ultrasound generators kicking out in excess of 150 decibels. That's equivalent to a jet engine at a distance of 30 meters. If you were hit by that level of sound at an audible frequency, it's safe to say you would need a new eardrum.

The Doctor's sonic screwdriver would actually need to be even louder still. Experimental ultrasonic hard drives are compact devices, with the ultrasound source positioned just a few thousandths of a millimeter away from the actual disks. But the sonic screwdriver often has to do the business from a range of a few tens of meters. Cranking out enough sound to do that could lead to one or two safety issues. Experiments have shown that ultrasonic vibrations exceeding 159 decibels begin to warm the human skin, and anything above 180 decibels is potentially lethal.

How then does the Doctor use the screwdriver at a distance without accidentally cooking his companions? A big step toward solving this problem would be to minimize the amount that the beam spreads out as it travels through the air. The more tightly focused the beam is, the less audible volume it loses with distance and so the less volume it needs to chuck out in the first place.

Technology to do this is currently under development. So-called parametric speakers are able to produce tightly directional, narrow beams of sound. They work on the principle that the higher you can make the frequency of a sound beam (or, equivalently, the lower the wavelength), the less it will spread out. "Dispersion of the sound depends on the sound's wavelength in comparison to the width of the surface that generates it. As the wavelength of the sound decreases, the surface forming the sound effectively gets wider in comparison, causing the sound to be emitted 'straighter,'" says James Friend, an ultrasonics researcher at Monash University in Victoria, Australia.

Parametric speakers exploit this property to use extremely high-frequency sound to project lower-frequency sound in a straight line over a distance. Have you ever heard two sounds together that are meant to have exactly the same pitch, but one of them is slightly out of tune? Rather than just the two notes, you actually hear what's called a "beat"—a kind of pulsating "wowing" sound, with a frequency given by the difference in pitch of the two sounds (see figure 6). Now imagine you had two high-frequency ultrasound signals, again ever-so-slightly out of tune. Although you wouldn't be able to hear the signals themselves—because they're ultrasound, which is too high-pitched for humans to hear—the beat between them has a lower frequency (remember, it's just the difference between the frequencies of the two beams), and by tweaking the frequencies you could arrange to hear that. Now imagine that your two signals are controlled by a computer, so that the frequency difference between them can be varied in real time to make the beat a time-varying audible sound—just like a normal speaker system.

This is how a parametric speaker works. "Modern versions project the two ultrasound signals out of the same speaker by combining the two signals into one input," says Friend. That

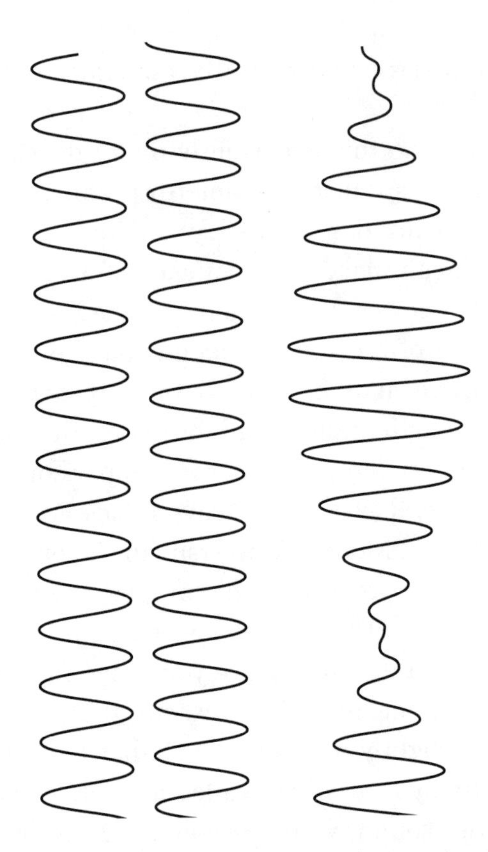

FIGURE 6. Two waves with slightly different frequencies (*left*) are added together to make a new waveform (*right*), the outer envelope of which is a wave with a frequency equal to the difference between the frequencies of the two added waves. The low-frequency "wowing" sound this makes is known as a "beat."

way the two signals can be projected along a single path, and the beat sound heard at any point along it.

But even with parametric speakers able to produce the ultrasound equivalent of a laser, the power requirements of a sonic screwdriver are still likely to be huge. The problem is getting enough of a grip on a screw head to tighten it. Undoing screws by hand is often hard work as it is—and swapping your trusty tempered steel for a spinning column of air certainly isn't going to make life any easier. Even if the screw head had fins on it to catch the air, the screwdriver is likely to suck up so much power that the device itself would have to be enormous—probably bigger than a living-room TV, certainly bigger than pocket-sized.

"To get a non-contact acoustic device to fit in the Doctor's pocket we would need to develop a new kind of power source," says Adams, "one with a very high energy density." Building such an energy source is, I dare say, a piece of cake for a Time Lord. But for you and me it probably means that handheld sonic screwdrivers will not be in stock at your local Home Depot any time soon—not until we're all carrying personal fusion generators in our pockets, anyway.

Locks and Land Mines

Of course, turning screws isn't all the sonic screwdriver is good for. In the show, it has a number of different settings—quite a large number in fact, if 2428-D (the setting used to weld barbed wire in "The Doctor Dances") is anything to go by.

Welding probably isn't the first application that springs to mind when dealing with sound waves, but ultrasound is already used for a similar purpose in manufacturing. "Every modern electronic part that looks like a black box with legs—such as CPUs, memory chips and so on—uses 'wire bonders,' which principally use high-frequency vibrations in the wires to solder them in place," says Friend. Welding is essentially the same process, but at a higher temperature.

On several occasions, such as in "The Pirate Planet" and "The Sun Makers," the sonic screwdriver has been used to open locks and crack safes. This presents a whole raft of new problems, such as projecting the ultrasound beam through the lock casing, and viewing the inside of the lock in the first place to see which parts of the mechanism to move, and in which direction. A gadget to do this would surely have to be automated. You can imagine low-power ultrasound beams being used to scan the inside of the lock and the images then being fed to a computer, which coordinates high-power beams to do the heavy lifting.

Alternatively, the screwdriver could just cut through the crucial components in the lock. Ultrasound is already used to cut small holes in leather, ceramics, glass, and semiconductors in manufacturing. It can even cut through metal via a technique called *cavitation,* in which high-frequency sound creates bubbles in a liquid coating the metal, which then collapse violently. "I once attended a demonstration where acoustic energy transmitted into fluids punctured and then disintegrated a beer can," says Friend.

If the sonic screwdriver can make a hole in metal, then giving it a good hard kick shouldn't be a problem either. So the Third Doctor wouldn't have had too much trouble detonating landmines with it, as he did in "The Sea Devils." But what about exploding a bottle of port? In the Ninth Doctor episode "World War Three," the Doctor grabs a bottle and threatens to use the sonic screwdriver to "triplicate the flammability" of the alcohol. He was only bluffing, but oddly enough there may have been an ounce of truth in his words. As Friend points out, it's possible to use acoustics to atomize a liquid into a fine spray—a process that can make the liquid more readily combustible (take gasoline vapor, for example, which is much more flammable than gas in its liquid form).

Although it returned in the Ninth and Tenth Doctor adventures, the sonic screwdriver was not popular with 1980s series producer John Nathan-Turner. He deemed it too much of an easy-get-out tool and ordered that it be written out. Consequently, the Doctor's favorite toy was destroyed by a Terileptil (a reptilian alien) in the 1982 adventure "The Visitation," prompting the Doctor to ruefully remark: "I feel like I've just lost an old friend."

6

Partners in Time

"What you need, Doctor, is someone to pass you your test tubes and to tell you how brilliant you are."

—Brigadier Lethbridge-Stewart, "Terror of the Autons"

It's a big, lonely Universe out there.

When his assistant Steven Taylor storms out of the Tardis in the First Doctor adventure "The Massacre," there's a touching moment as the Doctor laments the loss of all his companions. He briefly contemplates returning home to Gallifrey, but then remembers he cannot. With sadness he realizes how truly alone he is.

But not for long, of course. Steven soon returns, followed over the years by a succession of male, female, and even robot companions. The exact definition of "Doctor's companion" is something of a gray area—how many episodes and what level of involvement merits companion status? But most aficionados agree that by the end of 2005's season 27 (starring Christopher Eccleston as the Ninth Doctor), the Doctor has had 29 of them. Occasions in which he has appeared without one are rare and include the Fourth Doctor serial "The Deadly Assassin."

Why does the Doctor need a companion? According to Angela Carter, an occupational psychologist at the University of Sheffield, humans—and presumably Time Lords too—crave companionship for good reason. People are social animals and get depressed if they don't have other people around them, she says. But also humans evolved as part of a pack. As well as interacting socially with our peers, we lived for millions of years in an environment where we relied on them, for instance in hunting and gathering. You can hunt more effectively in a group, and you have others there to watch your back in case anything is hunting you.

Carter also thinks that companions help people to establish their sense of identity and clarify what their lives are really all about. "I learn from my peers what I do in my work, and in the same way we learn from our companions what our role is in life," she says.

The Doctor's companions have brought with them a host of quirks and odd abilities. The Sixth Doctor's assistant Mel has a photographic memory. In the adventure "Black Orchid," the Fifth Doctor's companion Nyssa meets her identical double (an occurrence that most biologists agree is nigh on impossible in unrelated people). Meanwhile, during "Horror of Fang Rock," the Fourth Doctor's assistant Leela changes her eye color from brown to blue. (It was blamed on the flash from an exploding spaceship causing "pigment dispersion." The real reason, however, was that the producers had made actress Louise Jameson wear red contact lenses up to that point to make her naturally blue eyes appear brown. But because the lenses limited her vision they eventually agreed that she could stop wearing them.)

Clearly, if you're a script writer, then having hapless assistants around to press buttons they shouldn't and get captured by the Cybermen is a handy plot device. Not to mention having them there to express questions that viewers are likely to have. But, let's be honest, there's another reason why *Doctor Who*'s producers chose to give the Doctor a companion—that is, a young, attractive female companion.

Earth Girls Are Easy

Rose Tyler fleeing the Autons may not be what fans are referring to when they talk about the show's bouncing wit. But it would be naive to suggest that glamour hasn't contributed in some way to making the series such a ratings success. "Men talk quite extensively about their early fantasies of latching on to one of the Doctor's assistants," says Petra Boynton, a sex and relationship psychologist at University College London. "Tom Baker's assistant Leela seems to come up a lot."

The science behind pornography is a bit baffling. What sort of evolutionary pressures make men fixate on images of females that they are unlikely ever to meet, let alone get the chance to mate with? It may be because television, and other media through which these images are delivered, are such modern inventions. TV's been with us for just 50 years or so, whereas evolution has spent millions of years training men to chase women across the African savannah. They've had it drummed into them that, as far as mating goes, seeing a female form in front of you is a pretty good start.

The counterexample, of course, is that pornography has a history that predates television by a long, long way, with rude images turning up on Greek pottery and even ancient cave paintings.

Imagery plays a large part in the male response to such images, but Boynton says there's more to it than that—there's a large fantasy element involved as well, with men often putting themselves into the role of the Doctor and imagining some kind of relationship with his assistant. *Doctor Who* is unique in this respect among science fiction shows—whereas, for example, *Star Trek* features overtly sexual female characters like Seven of Nine and T'Pol, the Doctor's companions are very much more "girls next door." And Boynton thinks this can make for an appealing fantasy. "It's a very attractive thing that you might meet this woman and she might be nice to you," says Boynton.

Her findings don't just apply to men. "We may have just as

many women who have over the years had a massive crush on Doctor Who, or one of the Doctor's assistants, or even some aliens," she says. "But probably not Daleks."

Nor is it confined to those of straight sexual orientation. Boynton has noted that a lot of lesbian teenagers have said they were drawn to female characters in the show. *Doctor Who* has had an enduring gay following, and perhaps this connection became more explicit while the show was under the guidance of award-winning gay writer Russell T. Davies.

One of Davies's most daring moves was the introduction of the show's first (overtly) non-heterosexual companion. Captain Jack Harkness is a conman and time traveler from the fifty-first century—where everyone, himself included, is bisexual. You might question how realistic this is, reasoning that evolution should favor heterosexuals in order to keep population numbers up. It does in the present day. But in the fifty-first century that could all change. As new technologies enable humans to transcend the limitations of biology, it may be that 3,000 years from now we'll no longer need to have sex to reproduce. "The imperative then will be sex for pleasure," says Boynton. "And any form of sexuality will be valid if you're just having it for pleasure."

Davies has Captain Jack flirt brazenly with the Doctor, culminating in the series' first gay kiss as the two say their farewells in *The Parting of the Ways*. It was a brave move for what's still a prime-time children's show. It certainly couldn't have been done back when the series was first launched in 1963, and probably couldn't even have happened 10 years ago. "When the show began, people had just about got their heads round being gay," says Boynton. "Even so, most people still thought it was wrong. Suggest to them the idea of someone being bisexual and it was: 'No! Good grief!'"

No Sex, We're Time Lords!

The gay kiss was as much a revelation about the Doctor's sexuality as anything else. Up until the 1996 television movie starring Paul McGann, the Doctor, despite a flow of nubile assistants, had remained staunchly asexual. (The only exceptions being his earlier life: the Doctor's very first assistant, Susan Foreman, is his granddaughter; and in the First Doctor story "The Aztecs" he is smitten by, and briefly engaged to, an Aztec woman called Cameca.)

Asexuality was fashionable among the British middle classes in the early twentieth century. "If you were a man of science or a man of God, you were not supposed to be troubled by the sex instinct," says Boynton. It's an attitude typified by Sir Arthur Conan Doyle's detective character from that era, Sherlock Holmes. And picturing the Doctor's early incarnations, you can imagine that this is exactly the kind of Doctor character that the writers were trying to portray—a distinguished eccentric, driven by science and his sense of decency. Is the emphasis now shifting toward more traditional "Hollywood" values?

It's certainly true that for years *Doctor Who* has flown in the face of other science fiction shows, where there's almost always a prominent love interest. The taboo was finally broken when Paul McGann's Eighth Doctor kissed companion Grace Holloway. The notion of the Doctor enjoying a less-than-platonic relationship with an assistant was not popular with some fans. So hats off to Russell T. Davies, who has taken the idea and run with it, deftly treading a line to give us a Ninth Doctor who's clearly in love with his assistant Rose, not averse to kissing men, and yet hugely popular with fans of all ages.

And what about his assistants? Rose Tyler exemplifies how the role of the Doctor's assistant has evolved over the years to mirror the perceived role of women in society. Back in the 1960s when *Doctor Who* began, her kind of assertive female character—who takes an active role and even gets the Doctor out of trouble now and again—wouldn't have made much sense to viewers. Early assistants were there to scream and look cute. Now we're fortunate enough to live in more liberated times, where it's okay for girls to punch aliens.

There's no doubting that the idea of lesbian companions would be popular with a certain viewing demographic—and probably the program-makers too. But Boynton has a better suggestion. It's one that's bound to ruffle feathers among fans but, if handled skillfully, could make the basis for some of the show's strongest human-interest storylines yet. "Isn't it about time," she says, "that we had a lady Doctor?"

Doctor Who does transgenderism. Parents: try explaining that one to the kids.

• Part Two •

ALIENS OF LONDON AND BEYOND

7

Other Worlds

"There are worlds out there where the sky is burning, and the sea's asleep, and the rivers dream; people made of smoke and cities made of song. Somewhere there's danger, somewhere there's injustice, somewhere else the tea's getting cold. Come on, Ace. We've got work to do."

—The Seventh Doctor to his companion Ace, "Survival"

Alien worlds are a staple of *Doctor Who.* Or at least they used to be. When the program-makers' budget gets tight, costly alien sets are often the first thing to go. That was the reason why Jon Pertwee's Third Doctor was stranded on Earth in his early adventures and why alien worlds were the exception rather than the norm during the adventures of the Ninth and Tenth Doctors.

Astronomers are finding alien planets around distant stars all the time. They use a variety of techniques—such as detecting the wobble of a star due to the gravitational tug of an orbiting planet, or measuring the dip in the star's brightness as a planet passes in front of it. As of January 2010, they had notched up over 420 discoveries. New projects promise to find even more, with the Holy Grail being to identify a faraway star with an Earth-like world orbiting around it. A planet like Earth—with an oxygen atmosphere and a temperate climate—would be a prime contender to harbor intelligent alien life. That's why NASA and the European Space Agency (ESA) are working on specially designed space telescopes that, orbiting high above the obscuring haze of our planet's atmosphere, will have the best chance yet of detecting one of these Earths away from Earth.

The Doctor's freedom to travel seems to depend on the existence of Earth-like worlds. After all, almost every planet he visits has a breathable atmosphere. How many times have we all wondered about the likelihood of that? Fred Taylor, a planetary scientist at the University of Oxford, doesn't think it's so unreasonable. He points out that plants give off oxygen as they photosynthesize—taking in carbon dioxide and sunlight to create the sugars they need to nourish themselves. "If the Doctor always lands on planets that have green plants, then the chances of there being oxygen there as well are pretty high," he says.

Taylor believes that there could be all sorts of other alien worlds out there—wildly different from the Earth and wildly different from anything we've seen so far in our Solar System. Statistically, it stands to reason. The Sun is just one of between 200 and 400 billion stars in our Milky Way galaxy. And the Milky Way is just one of millions of galaxies in a very, very large Universe. There could be literally billions of planets out there. So what are they all like?

The *Doctor Who* writers have thrown up some tantalizing possibilities—too many for an exhaustive discussion. Some are even too wacky to be debated within the realms of current scientific understanding. For example, there's the planet in "The Mind Robber" where things exist only if you believe in them. Then there's Zeta Minor, the Jekyll and Hyde world in "Planet of Evil" that turns hostile at night. Yet the writers have often hit on ideas for alien worlds that have resonated with astronomers.

Oceans of Acid

In the First Doctor story "The Keys of Marinus," the Doctor visits a planet that has seas of pure acid lapping over its surface. This is quite plausible. The planet Venus, Earth's nearest neighbor in the Solar System, has acidic clouds in its atmosphere. These are composed mainly of sulfuric acid, with traces of hydrochloric and hydrofluoric acid as well. If Venus wasn't so hot (surface temperature 450°C, thanks to a runaway greenhouse effect caused by the large amount of carbon dioxide in the atmosphere), it would be a simple matter for the clouds to condense into pools and perhaps even oceans on the planet's surface.

"Even in the Earth's atmosphere, clouds of sulfuric acid form as a result of volcanoes and man-made pollution," says Taylor. These clouds could condense if they were thick enough and didn't get washed away by ordinary rain first.

The oceans of Marinus are bordered by beaches of raw glass. This isn't a problem either. Glass is essentially sand that's been melted, fused, and resolidified. Terrestrial volcanic activity does this all the time—though typically it produces black glass, because it's so full of impurities. On the sea floor between Cuba and Mexico, geologists have found a stretch that's covered with volcanic glass, in the form of globes and spherules spat out during eruptions. Forming the glass on Marinus would require similar volcanic heat—the chemical "burning" effect of the acid in the oceans wouldn't do the trick.

"The Keys of Marinus" was an unusual *Doctor Who* adventure, because two of its episodes didn't feature the Doctor at all—actor William Hartnell was away on vacation. The Doctor himself needs some R and R now and again, and on several occasions planetary vacation destinations have featured in the show.

There's the vacation paradise world Florana in "Death to the Daleks," and the planet Argolis in "The Leisure Hive," which the Fourth Doctor and Romana choose to visit in preference to Brighton Beach. In "The Ribos Operation" we learn about the planet Halergan 3, a world consisting entirely of beaches—nothing but sea, sand, and palm trees. Beach worlds are an easy one. Fred Taylor thinks it's quite feasible for planets to exist composed purely of desert and liquid water oceans. "And you can lay out the continents in such a way that a large portion of the surface area is beach," he says.

Any Old Iron!

Planet Earth is mostly made of metal. Some 34 percent of the mass of the planet is accounted for by the heavy metal element iron. It's also made up from substantial amounts of magnesium, nickel, and aluminum. Its core region is around 7,000 kilometers across and made mostly of iron and nickel, although metallic deposits are also present in the mantle and in the crust, from where they can be mined.

But metals were nowhere near this abundant on the jungle world of Chloris, which Tom Baker's Fourth Doctor visited in "The Creature from the Pit." Chloris's crust had had all of the metal mined out of it. To mine every last scrap of metal from a planet's crust is, to put it mildly, quite an achievement—especially on such a manifestly undeveloped world as Chloris. The implication would have to be that there were only very limited quantities of it there in the first place. This might make some sense if Chloris resided far from its star. In the outer Solar System, most orbiting bodies are made from ice rather than rock and so naturally have a low metal content. But Chloris is clearly not one of these worlds. There is an abundance of rock and soil in which the planet's vegetation thrives, and the climate is obviously temperate—not the harsh, icy environment of a world far from its parent star. Chloris, it seems, is not crossing into science-fact any time soon.

In contrast to Chloris, and with more metal than the Earth, is a planet that featured in 1975's "Revenge of the Cybermen": Voga, a world made entirely of the precious metal gold. To Cybermen gold is toxic, which is why the cyborg race all but destroyed this world. Or so they believed—in "Revenge of the Cybermen," Voga's remains drift into the Solar System and are snagged by the gravity of the planet Jupiter.

Planets made entirely of a single metal aren't impossible. "If it was made of iron, I would have said 'yes, but with a few impurities,'" says Taylor. "In our own Solar System, the planet Mercury comes close to that." Iron is a common element in the Universe, but gold is much rarer. Taylor thinks it's unlikely there could be any kind of physical process that could gather together enough of it and then condense that material into something the size of a planet.

That said, he does know of a way to assemble large quantities of silver in space. And coincidentally it just happens to take place in Jupiter's atmosphere, right next to where Voga was meant to be in the series. Theoretical models predict that clouds of pure silver can form in the giant planet's atmosphere, at a level where the pressure is about 100 times the surface pressure on Earth. If these clouds were carried by updrafts to a height where the temperature is less than silver's boiling point, they could condense into droplets. "If you could think of a way to get that cloud out of there and condense it into a planet of pure silver then you could do it," says Taylor. "Something similar will apply for gold and other metals."

Metals in unusual states abound on other *Doctor Who* worlds as well. In "The Power of the Daleks," the Second Doctor arrives on the planet Vulcan, a world that has swamps with geysers of liquid mercury gushing into the sky. Interestingly, Vulcan was meant to be located between the planet Mercury and the Sun. In the nineteenth century, astronomers studying the behavior of the Solar System had hypothesized that there might exist just such a planet (and they even called it Vulcan). They'd observed curious behavior in the motion of Mercury around the Sun, which they thought could be explained by the gravitational pull of another planet, circling the Sun inside Mercury's orbit. But it wasn't to be. In 1915, Einstein published his general theory of relativity, a new theory of gravity that explained the orbit of Mercury naturally, with no need to invoke new planets (see Chapter 3).

Some astronomers still think there may be very small minor planets lurking between Mercury and the Sun—the rocky asteroids known as "vulcanoids." The difficulty with trying to detect them is that, being so close to the Sun, they are drowned out by its light. But that hasn't stopped astronomers from looking for them using telescopes in high-altitude jet aircraft, above air turbulence and the Sun's atmospheric glare. No positive detections have been made yet, but the search continues.

Of course, planet Vulcan in *Doctor Who* would be uninhabitable. Its temperature would be hotter than the surface of the planet Mercury, which itself is at 400°C. And geysers spraying liquid mercury metal into the atmosphere wouldn't be awfully healthy either. The element mercury is highly toxic, inducing symptoms such as tremors, dementia, and hallucinations. It's possible that liquid metals could flow freely on some planets. Venus, for example, has sinuous river beds visible in high-resolution radar images, stretching for hundreds of miles across the planet's surface. "We can't tell whether anything is still flowing in them, or what kind of fluid it is that cut them," says Taylor. "But candidates could be low-melting-point metals, such as lead or tin."

Venusian Karate

The planet Venus was especially close to the heart(s) of the Third Doctor, who extolled the joys of Venusian hopscotch, enthused over the soporific properties of Venusian lullabies, and even battled his foes with various forms of Venusian martial arts.

As we've seen, Venus isn't the most idyllic place to be. It's searingly hot, the atmosphere is unbreathable carbon dioxide laced with clouds of acid, and the surface pressure is about 90 times what it is on Earth. But that hasn't stopped some scientists speculating on the possibility of life there. Not on Venus's surface—but about 50 kilometers up, where the temperature drops to 70°C and the pressure is equivalent to that at Earth's surface. Still a bit warm, but reasonably hospitable. In fact, if you had an oxygen mask to help you breathe, this is perhaps the only place in the Solar System beyond the Earth where you wouldn't need a spacesuit to protect you.

Scientists at the University of Texas in El Paso say that life could exist at this altitude and that it could explain some observations of Venus's atmosphere that otherwise don't seem to add up. For example, readings from space missions to the planet have found hydrogen sulfide and sulfur dioxide in the atmosphere. These gases usually react with one another, to form other compounds. So the fact that they are found together in an unreacted state on Venus suggests that something on the planet is continuously replenishing them. Life is one possible source. More intriguing still is the existence of carbonyl sulfide in the atmosphere, believed by some to be a clear indicator of biological activity.

Venus hasn't always been such a heat trap, and it may be that life began on the planet long ago when it was cooler and more Earth-like. The Texas team suggest that as the greenhouse effect began to warm Venus, life retreated to the upper atmosphere. Perhaps what we are seeing today are the final vestiges of that life, clinging on as the planet below bakes. We may soon find out. At the time of writing (October 2009) the European Space Agency's Venus Express spaceprobe is in orbit around the planet, studying its surface and atmosphere as part of a mission that will last until 2012.

Venus is the only planet in our Solar System to spin in "retrograde," that is, it spins clockwise on its axis while it orbits the Sun in a counterclockwise sense. Those who know their Solar System science may argue that Uranus also spins in retrograde. However, Uranus's rotation axis is at a whopping 90° to its orbital axis, to within a degree. And so, to all intents and purposes, Uranus rolls around the Solar System "on its side"—very different to Venus, which is tilted just 3.4° from upright. In

"Meglos," the Fourth Doctor visits Tigella, another planet that has this rare property. Scientists think that Venus's retrograde rotation is down to internal friction, or turbulence, or even the effects of "chaos." (Chaos is a branch of mathematics dealing with the behavior of extremely sensitive systems. You've probably heard of the "butterfly effect"—the idea that a butterfly beating its wings in China could trigger a tornado in Texas. That's chaos theory applied to the weather. The state of the weather three days from now is so sensitive to the state of the weather—and butterflies—today, that predicting it with any accuracy is nigh on impossible.)

If you think Venus spinning backward is strange, then imagine living on a planet where the rotation periodically switches from prograde (spinning in the same sense as it orbits) to retrograde and back again. Just to make things more interesting, the length of day and night constantly vary. And sometimes there's no night at all. This is what life might be like if you were on a planet orbiting in a binary star system—a system comprising two stars, each orbiting around the other. The Doctor has encountered such worlds, for example the planet where the Gonds live in "The Krotons."

"If you're looking for Earth-like planets that are warm and wet, then you're typically looking for a single-star system," says Taylor. "But that doesn't mean that there aren't planets around multiple stars."

Many planetary orbits in a system of two or more stars are unstable, leading to the planets being flung out into space or transferred onto a different orbit altogether. One solution is to look for planets orbiting so far out that, from a gravitational perspective, the stars in the system all appear as one. (In the same way that two stars in a binary system a long way away will appear visually as one—unless you look at them through a powerful enough telescope—so their combined gravitational field looks more like that of a single star the further away you get.) But this is bad news if you're looking for life because, that far out, conditions will be too cold for anything to survive.

There are, however, figure-eight-shaped orbits meandering between double stars that are stable, although the jury is still out as to whether planets on such orbits could really sustain life. “You get a strange climate on planets like this,” says Taylor. “We don’t really know what kinds of life are possible or what conditions it can tolerate.” Until we do, it’s impossible to say whether the Gond world would be a haven for life or a barren husk completely devoid of it.

Ice and Fire

A planet’s orbit can have a huge impact on its climate. For example, look at the bizarre cycle of summers and winters that the Fourth Doctor discovered on the planet Ribos when he visited it in the 1978 adventure “The Ribos Operation.” Ribos has a highly elliptical orbit that takes it in close to its star and then far away, out into deep space, setting up a cycle of alternating hot and cold eras, known as “sun-time” and “ice-time,” each lasting 32 years.

This is broadly possible, although the ellipticity of Ribos’s orbit would need to be small. Otherwise, the variation in temperature between the warm and cold eras could be colossal, rendering the climate totally uninhabitable.

The Earth actually follows a similar orbit to Ribos. Earth’s path around the Sun is mildly elliptical—nowhere near as much as Ribos’s, but enough to bring about a 5°C temperature difference between its closest and farthest approaches to the Sun. Ironically, the closest approach takes place in January, and so if orbital distance was the only factor regulating the planet’s temperature, the entire planet would be celebrating New Year’s Day on the beach—rather than just the southern hemisphere, as is actually the case.

The reason only the south gets a warm New Year’s is that there’s a much stronger effect at work: the Earth’s tilt. Earth’s rotation axis is tipped over at 23.5° to its orbital axis. This means that throughout the year each hemisphere takes its turn to tip

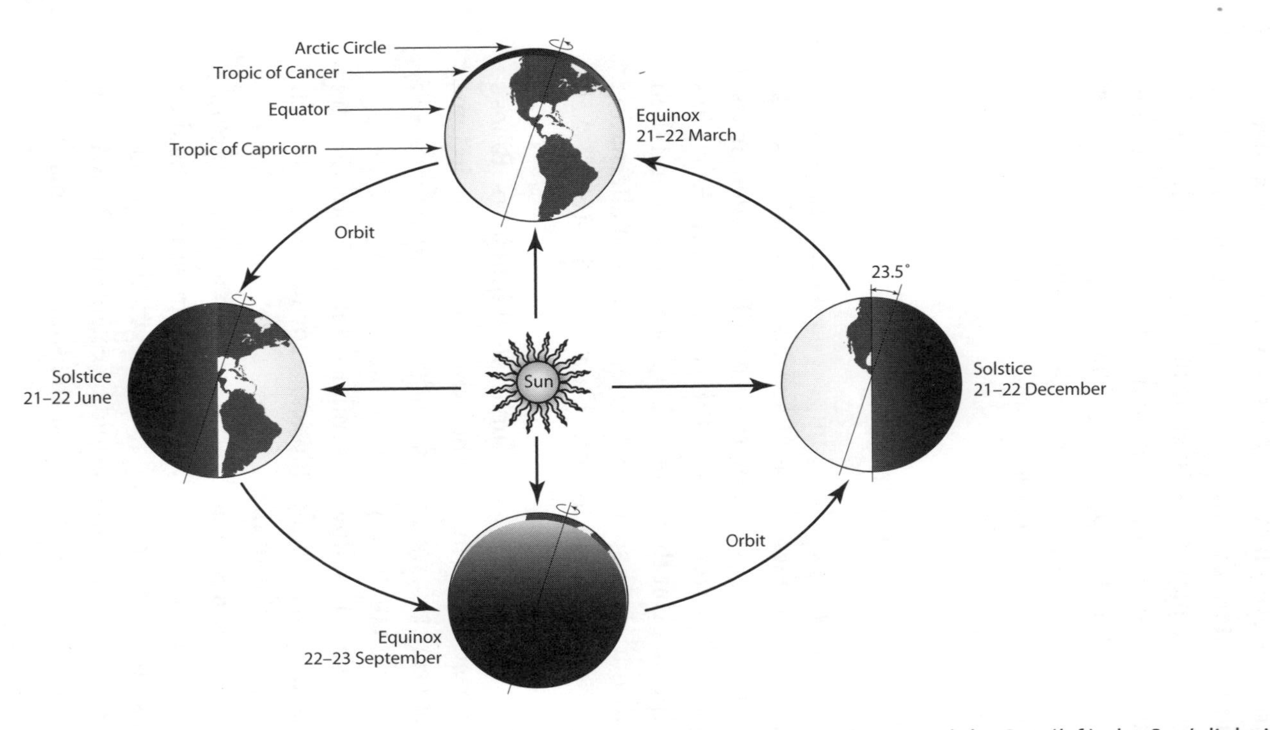

FIGURE 7. How the Earth's tilt is responsible for the seasons. When the northern hemisphere tilts toward the Sun (*left*), the Sun's light is spread over a smaller area than when it's tilting away (*right*), giving rise to summer and winter, respectively. This also explains why summer in the northern hemisphere coincides with winter in the southern hemisphere and vice versa.

toward the Sun or away from it. When a hemisphere tips toward our star, sunlight (and heat) falling on it is concentrated over a small area—it's summer in that hemisphere. When it tips away, the same amount of light is spread out over a larger area, making it cooler—and that's winter (see figure 7). At the latitude of the UK, that amounts to a temperature difference between summer and winter of about 30°C, which easily drowns out the effect of orbital distance. It happens that at the time of New Year's Day, it's summer in the southern hemisphere and winter in the north.

Before we leave the subject of "cold," the Third Doctor serial "Planet of the Daleks" featured a world called Spiridon where from time to time sub-surface pressure would force slushy ice up through the planet's crust, from where it would erupt like a volcano, smothering the surrounding jungle. In fact, ice volcanoes have been found for real on Triton, the largest moon of the planet Neptune. When the *Voyager 2* spacecraft flew past the moon in 1989, it photographed a plume of slushy material (probably a mixture of nitrogen, dust, and methane compounds) rising 8 kilometers above the surface.

The icy eruptions are thought to be caused by the Sun melting volatile materials beneath Triton's surface during the moon's summer. There's mounting evidence that so-called cryovolcanism also takes place on other worlds, including Jupiter's moons Ganymede and Europa, and the nearby planet Mars. In 2005, reports emerged that Saturn's moons Enceladus and Titan also play host to these amazing icy eruptions.

Northern Star

"Lots of planets have a north." That's what the Ninth Doctor says to his new companion Rose when she asks him why, if he claims to be an alien, does he sound like he's from northern England? The statement is remarkably accurate—lots of planets really do have a north. In fact, all of the planets in our Solar System, with the exception of Venus, have one—that is, a planet-wide magnetic field, arcing between a south pole and a north pole.

Planetary magnetic fields are thought to be produced by swirling currents of molten metal inside a planet's spinning outer core. In the same way that a generator creates an electrical current, the conducting liquid metal in the core sets up a current that in turn generates a magnetic field.

The exact location of "the north," however, isn't fixed. Earth's magnetic north pole wanders by as much as several miles every year. And every 250,000 years the field undergoes a complete reversal, where north and south spontaneously swap places. Although no field reversals have taken place in modern times, scientists discovered the effect by studying the magnetic orientation of solidified lava from different layers on the ocean floor, revealing the planet's magnetic history. The reason for the reversals isn't yet known.

Magnetic fields are key to the emergence of life on a planet, deflecting away the high-energy charged particles that are a major component of the harmful radiation emitted by the Sun.

Despite his accent, the Ninth Doctor isn't from the north at all—at least not on Earth. No discussion of the amazing worlds of the Whoniverse would be complete without mentioning the Time Lord home planet of Gallifrey. This yellowy world lies 250 million light-years away from Earth. That distance is well outside our Milky Way galaxy, but strangely enough coincides with the best estimate that astronomers have for the distance to a gravitational anomaly known as the "Great Attractor."

Located at the heart of our local supercluster (the cluster of galaxy clusters that we are part of), this invisible concentration of matter was detected by the effect it has on the motions of other celestial objects. It's estimated to have a mass equal to tens of thousands of galaxies. A giant black hole is one possible explanation. Interestingly (with tongue planted firmly in cheek), *Who* fans will remember that Gallifrey is home to the Eye of Harmony—a black hole created by the Time Lord engineer Omega, which is the source of the Gallifreyans' power to travel through time.

But Gallifrey was an ill-fated world. In the nine-year hiatus between 1996's TV movie and the series's 2005 relaunch, the Doctor has been away fighting the last great Time War against the Daleks. Casualties on both sides were heavy. The Daleks were (almost) wiped out. The Time Lords, with the exception of the Doctor, were obliterated. And, worst of all, Gallifrey was destroyed.

But just how easy is it to destroy a whole planet? It's a simple calculation to find out. All you need to know is the planet's "gravitational binding energy." That's the energy locked up in the gravitational field holding the planet together—you can work it out to a good approximation from Newton's laws of gravity. If an attacking Dalek fleet is able to deploy a weapon with an energy yield equal to or greater than Gallifrey's binding energy, it'll blow the planet apart.

Not surprisingly, it takes an awful lot of energy to do this. Let's assume Gallifrey is similar in size and mass to the Earth. The binding energy of an Earth-like planet is around 10^{32} joules (J), a unit of energy that physicists use. Compare that to the explosive power of the largest atomic bombs. These release energy equivalent to detonating around 50 megatons of TNT, about 10^{17} J—staggeringly destructive, yet still a million billion times too feeble to do for a planet.

The kind of destructive power that the Daleks actually need is so stupendous that it would take our entire Sun nine days of continuous shining to churn it all out. What kind of weapon could muster that much power, let alone deliver it all in a heartbeat? The most efficient explosive we know of is antimatter. A particle of antimatter is the complete opposite to a particle of normal matter. That means that if you take a particle of normal matter and collide it with its antimatter counterpart the two will annihilate, their entire combined mass being turned into energy. We can use Einstein's famous formula $E = mc^2$, relating energy (E) and mass (m) by the speed of light (c), to work out how much antimatter the Daleks require. It turns out they'll need a bomb weighing around 10^{15} kilograms, about the mass of Mars's moon Deimos.

That's a heck of a lot of antimatter. Physicists on Earth today typically make the stuff a particle at a time, and it takes huge amounts of energy to do it—millions of times more than is given off when the antimatter eventually annihilates. So for the time being it looks as if Gallifrey is safe.

Even so, I fancy all this talk of planetary devastation has got just a little bit depressing—what we need is something to lighten the mood. How about a nice trip to the zoo? Forget monkeys and elephants, though—this is a zoo with a difference. We're off to see energy beings, shape-shifters, and little green men.

And for your own safety, please do not feed the Daleks.

8

Carnival of Monsters

"The government are gathering together all the experts in aliens, and who's the biggest expert of the lot?"
"Patrick Moore?"

—The Ninth Doctor and Rose, "Aliens of London"

If there are billions of planets out there in space, as we saw in the last chapter, then the next thing we want to know is: could there be life on any of them? In *Doctor Who*, there's certainly plenty of life in the Universe. From the genetically engineered cyborgs that are the Daleks to the shape-shifting Rutans and the plant-like Vervoids, the Whoniverse is teeming with creatures of varied form and demeanor.

In our Milky Way, the reality is less clear. We know for sure that Earth-like worlds can sustain life—because we have one very good example. And so if there are Earth-like worlds out there, we can say it's at least very plausible that there will be life on them.

It's unlikely though, that any life on such worlds will look anything like human beings. The evolution of life is a process that scientists call a "random walk"—it's shaped by chance events and random dice rolls that make it impossible to predict. The fact that we have two arms and two legs, that our airway crosses our food way (placing our nose above our mouth) and that our eyes are above our nose is all completely arbitrary. Even if we could rewind the tape on Earth and run the evolution of life here all over again, the chances of us arriving at anything resembling a human being are absolutely minuscule. Jack Cohen, a biologist at the University of Warwick in England who has studied the possibilities for alien life, thinks that even vertebrates—creatures with a spine—would be unlikely to appear second time around.

"A lot of the *Doctor Who* aliens are anthropomorphic, and they just wouldn't happen," says Cohen. "You can't get a creature that looks like a person unless it's a descendant of that very fish that came out of the water on Earth."

That said, there would be a degree of what biologists refer to as *convergent evolution.* During the evolution of life on Earth, certain biological traits have appeared over and over again—independently. For example, flight has emerged on three separate occasions, in insects, birds, and bats. And eyes have arisen no fewer than 40 times. It seems that in the course of Earth's history, whenever life in different habitats has faced a problem of some sort—that is, how to get from one tree to another or how to sense the immediate surroundings—in each case nature has solved the problem in remarkably similar ways.

It seems natural to extend this principle of convergence beyond the Earth and to life on other worlds as well. And so on a distant Earth-like planet, although we can't expect to find creatures that are manifestly humanoid, we can expect a few similarities. We can expect to see creatures that have limbs with digits on the end so they can manipulate their environment. We can expect them to have eyes. There are likely to be green, photosynthetic plants there. And in that case it seems likely that the life-forms will have evolved lungs of some sort, to breathe the oxygen that all these plants will be producing. It also seems reasonable to expect at least some of the creatures on such a world to be intelligent.

In the 1960s, American astronomer Frank Drake attempted

to estimate the number of intelligent civilizations in our galaxy—a number known simply as "N." He wrote down an equation, expressing N in terms of various factors such as the number of habitable planets, the formation rate of Sun-like stars, the average lifetime of a communicating civilization, and so on. The trouble is that scientists' best guesses for the values of these factors vary so wildly that the resulting value of N can be anywhere in the range of two ten-millionths to many millions. If we take the first figure as gospel, this is going to be a rather short chapter. So let's live a little. Let's interpret the upper bound as a demonstration that a galaxy populated by a large number of intelligent civilizations isn't an impossible concept. In that case, what can we say about them?

For a start, would they want to meet us? Or more to the point, is it right for us to want to meet them? Contact with another race from across the galaxy would completely overturn the natural course of progress on a planet. Is that ethical? The Time Lords certainly think not and have ruled against it. They hold to a policy of "non-interference" in the affairs and events of the galaxy.

It's a code of conduct that the Doctor himself rarely adheres to. In fact, he seems to relish breaking it. "Interfere! Of course we should interfere—always do what you're best at, that's what I say," he pronounces in the Fourth Doctor serial "Nightmare of Eden." And it's the charge of violating this key Time Lord policy (not to mention also breaking the first law of time—see Chapter 3) that the Doctor is summoned back to Gallifrey to answer in the 1986 series of linked adventures, "The Trial of a Time Lord."

Seth Shostak, a senior astronomer at the Search for Extraterrestrial Intelligence (SETI) Institute in Mountain View, California, thinks that real space-faring civilizations are highly unlikely to be this considerate. He points out that—ethical or not—most of the human exploration of the Earth so far has been conducted with interference as a key motive. We've set out to discover new peoples and study them, trade with them, convert their religious beliefs, and occasionally bring them back to do our chores. Even when the idea was to "do no harm"—as were the British Admiralty's orders to Captain James Cook—he inevitably and irreversibly changed the South Sea islander communities that he discovered.

"This all falls in the realm of 'alien sociology,' and we don't know a heck of a lot about that," admits Shostak. "But based on the sociology we do know, I think you can count on lots of interference."

Which is all good news for the Doctor. So now all he needs are some alien races to go and interfere with.

Galactic Zoology

Earth-like worlds are one thing, but what about other sorts of planets? Could there be life on scorched volcanic worlds? Or frozen planets of ice? Or on other bizarre sorts of worlds the likes of which we've never imagined?

Jack Cohen thinks there are likely to be so many life-forms in our galaxy that cataloguing and classifying them all would be a monumental task and maybe not even possible. To get an idea why, picture a piece of turf in a typical English churchyard, say a meter square and 15 centimeters deep. Just in that small chunk of soil there are an estimated 30 species of insects, 450 species of nematode worms, and 1,500 and maybe more species of bacteria. And of those, biologists have named and classified maybe 25 of the insects, 50 of the worms, and just a tiny fraction of the bacteria. We've hardly scratched the surface in cataloguing life on Earth.

"We've labeled nearly a million species of insect. There's a guess that there's actually three times that number," says Cohen. "There may be ten times that." If we don't even know all the species in our own backyard, then how would we possibly cope with a whole galaxy's worth? Perhaps we wouldn't.

But it's not the alien bacteria and nematodes that the Doctor comes to see. He's here for the big stuff, and that's what you and

I are interested in too. So what are we likely to find? And, more to the point, what exactly is it going to look like?

There are a number of environmental factors that will influence the form of an alien species. Foremost is gravity, which determines whether aliens are short and squat, or tall and spindly. Gravity also determines how dense a world's atmosphere is, and this in turn fixes the lung capacity of life-forms (assuming they do actually breathe the atmosphere). Creatures on a low-gravity world with a thin, rarefied atmosphere would need large chest cavities with very well-developed lungs to extract a useful amount of oxygen (or whatever chemical they breathe). Conversely, if gravity was high, then the atmosphere would be dense and squeezed to high pressure, allowing lungs to be very much smaller. The humidity, temperature, and composition of the atmosphere would also help determine the precise form that life takes.

Life-forms must also possess a number of basic senses if they are to survive, such as the abilities to see, hear, and smell—each of which will have a dedicated organ, probably situated close to the creature's brain.

The senses of an alien will be adapted to the sort of planetary environment that it's living in. Take bats on Earth, for instance. They hunt at night in the dark and so have little use for eyesight. Instead they use echolocation to see—emitting sound waves and listening for the echo from nearby objects. A bat's echolocation is so sophisticated that it can identify a tasty insect simply from the pattern of the sound that bounces back from it. Cetaceans, such as whales and dolphins, also use this technique. Again, it suits their particular environment. Because water is a denser medium than air, sound waves have an extremely long range in the ocean—whale songs can be heard up to 3,000 kilometers away.

Other more exotic aliens may have different kinds of senses completely. Sharks, for example, and some other fish use electro-reception to pick up the electrical signals given off by other creatures in the water.

In addition to eyes and ears, a successful species will also have a sense of taste, so it can tell what's worth eating and what could be poisonous, and a sense of touch, so it can manipulate its surroundings, and feel and react to potential hazards such as temperature changes.

Ice Worlds

It's possible that there are creatures surviving on planets where the temperatures are extreme. On Earth, so-called hyperthermophilic bacteria have been seen thriving after 24 hours in an autoclave oven at 121°C. At the other end of the spectrum, the psychrophilic (cold-loving) bacterium *Psychrobacter cryopegella* can exist in Siberian permafrost at temperatures dipping down to −20°C.

The Cryons, in the Sixth Doctor adventure "Attack of the Cybermen," were a humanoid psychrophilic species. The native inhabitants of the planet Telos, they couldn't survive in temperatures above the freezing point of water (0°C). The Cryons were all but wiped out by the Cybermen when the evil cyborg race seized Telos as their own.

Another species that would have to live in low temperatures are the Kastrians from the 1976 adventure "The Hand of Fear." That's because they were a race of creatures based upon the chemical element silicon. Silicon shares many of the chemical properties of carbon, on which life here on Earth is based, leading some biologists to suggest that it could form the basis for an alternative biochemistry. Like carbon, silicon is able to form what are called *tetravalent bonds* with other atoms and molecules (see Chapter 11 for more on this), enabling it to form complex molecular structures.

But carbon or silicon alone is not enough. Life also requires a solvent in which the base chemicals are dissolved. In the case of our carbon-based biochemistry, this job is done very well by water. However, silicon compounds are not stable in water and, at the temperatures for which water is liquid, silicon dioxide (the silicon equivalent of carbon dioxide, which silicon-based animals would breathe out) is a solid: sand.

The solution is to use ammonia as a solvent instead. Yet this exists in the liquid state only between –75 and –34°C. "At this temperature, the chemistry is very, very slow," says André Brack of the Center for Molecular Biophysics in Orléans, France. "But that doesn't mean it's something to be excluded."

Nevertheless, it does make it unlikely that any silicon life we encounter will be anything more than bacteria. Life on Earth began some 3.5 billion years before our current era, but primitive multicellular creatures—the beginnings of larger life-forms like ourselves—arose only about 700 million years ago. Given that silicon chemistry is even slower than this, it does raise doubts as to whether higher forms of silicon-based life have had time to evolve yet within our galaxy.

Clearly in the *Doctor Who* universe they have, although their need for a cold climate wasn't realized by the show's writers. In "The Hand of Fear" we're told that the Kastrians' home planet used to be Earth-like but was rendered "uninhabitable" when solar winds reduced it to an icy wasteland. Oops.

Believe it or not, there are creatures on Earth that are partly made from silicon. "Diatoms" are a single-celled species of plankton, each encased in a silicon-based shell.

Finest Green

Probably the most stereotypical alien attribute in science fiction is green skin. And *Doctor Who* has paid homage to this, most notably with the Swampies in the Fourth Doctor story "The Power of Kroll."

Green skin isn't actually as exotic as we might think. Lots of smaller creatures on Earth, such as chameleons, have it. And a few larger ones do as well, like crocodiles. There's even a species of sloth that has green fur—although that's a bit of a cheat. The

Amazonian three-toed sloth has a strain of blue-green algae growing in its fur, which provides it with both camouflage and nutrition.

Why would an alien have green skin? One motivation would be nourishment—any life-form hoping to feed itself partially or wholly by photosynthesis would also need to have permanent green pigment in its skin. The actors playing the Swampies in "The Power of Kroll" must have wondered if they were going to have permanently green skin too, when the green paint they were wearing proved exceptionally difficult to remove. The cast were forced to call in on an American airbase near the filming location and use industrial solvents to get the stuff off. (For more on the pros and cons of green skin, see the discussion on mobile plants in Chapter 16.)

It was on another swamp world that the Doctor encountered creatures capable of a rather more unusual alien ability. The inhabitants of the planet Alzarius are able to heal injuries remarkably rapidly. For instance, when the Alzarian Adric (who decides to remain with the Doctor as an assistant at the end of the adventure) cuts his knee open on a rock, it heals up within a few hours. This ability is common in lower life-forms on Earth such as flatworms, which are able to heal major wounds within an hour or so, though this isn't something you're going to see more advanced life-forms doing any time soon. Higher creatures—like humans—take longer to heal because their cells are more strongly differentiated into different types of tissue, such as skin or heart or brain. When flatworms undergo rapid healing their cells actually revert, or de-differentiate, back into a more generic "stem cell" state, from where it's easy to grow new tissue. This sort of reversion isn't possible in higher mammals.

The Alzarians also evolve extremely quickly. Their species has evolved from primitive marsh creatures to the current human-like form in just 4,000 generations. Compare that to the estimated hundreds of thousands of generations separating today's humans from the ancestor they share with modern chimps.

This actually fits with a theory called "punctuated equilibrium," developed in the early 1970s by paleontologist Niles Eldredge and late evolutionary biologist Stephen Jay Gould. It says that evolution isn't a steady process but happens in a series of fits and starts (punctuations) with large periods of little change in between (the equilibria).

In June 2005, a team of geneticists from the University of Washington, in Seattle, found that most of the changes in the human DNA molecule known as chromosome 2 did indeed occur in a narrow window of time between 10 and 20 million years ago, supporting the punctuated equilibrium hypothesis.

Maybe the Alzarian race are right in the middle of just such a phase of rapid evolution.

Morphing Monsters

Shape-shifting is another popular sci-fi theme that's been explored by the writers of *Doctor Who*. In "Horror of Fang Rock," the Fourth Doctor serial from 1977, we meet the Rutans—the mortal enemies of the Sontarans (see Chapter 14). The Rutans are able to change their appearance at will, assuming any form they wish (just as well for them, since in their natural state they resemble glowing cabbages with tentacles).

Shape-shifting in this way must be great fun, but it doesn't sit well with the laws of physics. The big problem is the basic rule that matter has to be conserved. And if it's not conserved, then you need a plausible explanation for where surplus matter is going to or where extra matter is being piped in from. For example, you can't have a shape-shifting mouse suddenly turn itself into an elephant without some kind of explanation for where all the extra mass it needs to do this has come from. Or where all that mass goes again when it turns itself back into a mouse.

So unless there's some kind of spacetime wormhole (see Chapter 3) linking each Rutan warrior to a repository of organic material, there seems to be little hope that this could ever actually work. Unless, of course, the Rutans concede to shift only into creatures that are made of exactly the same amount of matter as they are. Even then, the problem of how to change color, texture, and form seems way beyond the bounds of biology as we currently understand it.

"I don't believe it a bit, and neither do you!" concludes biologist Jack Cohen.

If changing form seems outrageous, then creatures that have no physical form whatsoever must be totally out of the question. The Vardans were a race of beings made of pure energy, encountered by the Doctor in "The Invasion of Time." They were able to transport themselves along telecommunication broadcasts and even hitch a ride on the Tardis's scanner beam. Physicists are quick to attack the idea of energy beings. They say that radiant energy must travel at the speed of light, and so any beings made of it must also travel at the speed of light, which can't be very useful when they want to stop for a chat.

Luckily for the Vardans' social lives, Cohen thinks he might have the answer. A living thing is not an inanimate lump of matter like a rock. He argues that it's more like a fountain or a whirlpool, in which the matter isn't the important thing—it's what the matter's doing that counts. Living creatures are made from matter that's in a constant state of flux—new cells being made as old ones die off. "The living thing is something like the energy involved in this process," says Cohen. "It's certainly not the matter because the matter is changing." And so he envisages a kind of life-form that goes through this same energy process, but does it in a way that has dispensed with matter entirely.

If there are creatures out there in our Universe anything like the Vardans, then perhaps that's roughly how they work. If so, the exact details of their biology are still way beyond our understanding. How (and what) do they eat? With no physical bodies, how do they interact with their environment? How do they communicate? How do they reproduce? The Vardans may seem unlikely when considered in terms of today's science, but as is so often the case with far-reaching ideas, we rarely get to see the full, big picture in the first glimpse.

That was certainly true when the show's writers came up with the idea for what are now one of the Doctor's most renowned enemies. Who'd have thought when they debuted in 1966 that men in silver jumpsuits with pipes on their heads could cause a Time Lord so much trouble?

9

The Cybermen

"UNIT's been very busy, Doctor. We've also got high-explosive rounds for Yetis and very efficient armor-piercing rounds for robots. And we've even got gold-tipped bullets for you-know-what."

—Brigadier Lethbridge-Stewart, "Battlefield"

Within a few years of *Doctor Who*'s launch in 1963, the show's writers and editors had bumped into a bit of a problem. They'd realized that writing good science fiction isn't simply a case of making it all up as you go along. In particular, the fact that none of them knew very much about science was starting to show through in their stories. But script editor Gerry Davis had the solution. He decided to appoint a science adviser to keep the writers up to speed with the latest scientific developments. The name of the person he appointed was Kit Pedler. Head of the Electron Microscopy Department at University College London, Pedler had also been working in television on the BBC science documentary series *Horizon*. He was soon spotted by Davis and recruited onto *Doctor Who* to give story-lines a stronger scientific base. It wasn't long before Pedler proved his worth. One of his very first creations has proved to be one of the Doctor's most enduring adversaries: the Cybermen.

In 1954, at Peter Brigham Hospital in Boston, Joseph E. Murray had performed the world's first successful kidney transplant. Doctors followed this up in the 1960s by successfully transplanting lungs and livers from newly deceased donors into living patients.

This work fascinated Pedler, and he wondered how far the technology could be taken. He envisaged a time when we would be having so much transplant surgery that little of our original bodies would remain. The result of his musings was a race of creatures who were once humanoid but, having experimented with cybernetic modifications to their bodies, were now almost entirely robotic. With their organic body parts went much of their capacity for emotion—the Cybermen are a cold, logical race, capable of the most callous acts in order to achieve their objectives.

The Cybermen made their debut in the 1966 First Doctor story "The Tenth Planet." The Doctor's encounter with them was so traumatic that the adventure ended with his regeneration. They return with a vengeance in the Tenth Doctor adventure "Rise of the Cybermen." But what's the reality of cybernetic organisms, or "cyborgs" as they're known? Was Pedler's vision correct—are we going to become a species of man-machine hybrids? And what will it mean for us if we do?

Chip in Your Shoulder

Some might argue that a blind man using a cane to "see" is a cyborg of sorts—he's using a mechanical aid to boost his body's natural abilities. And if that's the case, then we are already a planet of Cybermen. Most of us, however, would view a cyborg as a creature with one or more electrical or mechanical components embedded in its body. These too are now starting to become surprisingly commonplace. Pacemakers to maintain the heart's rhythm, cochlear implants to improve hearing, even hip replacements are all examples of artificial systems being used to enhance the human organism.

In 1998, Kevin Warwick at the University of Reading near London decided to take this a step further by becoming the first human being to have a microchip, measuring 23 by 3 millimeters, surgically embedded in his left arm. As he walked through the university's Department of Cybernetics, sensors would detect the chip and open doors and switch on lights as he approached. His computer would greet him before he'd even touched the keyboard.

Warwick's "Project Cyborg" lasted 9 days before he had the chip extracted—mainly to avoid any potential medical complications. He says that in this short space of time he quickly regarded the implant as being part of his own body. Patients implanted with pacemakers and cochlear implants report similar rapid acceptance of the devices.

Chips like this are already finding real-world applications. The Baja Beach Clubs in Rotterdam and Barcelona now require members to have microchip implants in order to access certain exclusive areas. And you don't need to pay by cash or credit card—your bar bill is charged directly to your account, using the chip as an identifier. "They claim they have a waiting list of people who want the implant," says Warwick.

In 2002, Warwick upped the game again. He had another chip implanted, also into his left arm, but this time it was directly linked to his nervous system. Through the device, Warwick was able to transmit signals to his computer simply by wiggling his fingers. The signals could then be played back and received by the chip in an attempt to replicate the same movements. He also found that emotional responses, such as shock, could be picked up by the chip, recorded on his computer and played back in the same way.

Next, Warwick's wife had a similar chip implanted in her own median nerve. Using the Internet, the two were able to send signals directly to each other's nervous systems. Warwick even used his implant to control, from his lab in Reading, a robot hand at New York City's Columbia University via the Internet.

"My nervous system didn't stop at my body's limits," he says, "but rather where the Internet link concluded."

The American Medical Association is now looking at the possibility of implanting chips in patients in order to access their medical records. The future potential medical applications of cybernetic technology are mind-boggling. International medical technology firm Medtronic is now producing cybernetic brain implants that alleviate the symptoms of Parkinson's disease. The implants deliver electrical stimulation to the areas of the brain that control movement and muscle function, blocking the signals that are responsible for the tremors experienced by Parkinson's sufferers. Warwick has seen the device in action. "One second the person can't move—they can't get up—and the next second it's as if they don't have Parkinson's," he says.

And if you think that's incredible, there have been a number of recent cases of patients being given brain implants that allow them to control their computers by the power of thought alone. Matthew Nagle was rendered quadriplegic during a knife attack in 2001. In 2004, a tiny chip in his head called a BrainGate, was implanted by a team at New England Sinai Hospital, in Stoughton, Massachusetts. Using it, he was able to move the cursor around on his computer screen—despite being paralyzed from the neck down.

Once the chip was in place, Nagle was asked to think about moving his cursor up, down, left, and right. Then his computer was programmed to recognize the pattern of signals that the chip picked up from his brain in each case and move the cursor accordingly.

Before long, he was not only able to change the channels on his TV and adjust the volume simply by thinking—he soon achieved sufficient control to play the game Tetris as well. Could this kind of technology ultimately enable paralysis victims to regain control of their limbs via artificial nerves? Could it even allow amputees to control prosthetic robot limbs directly from their brains? In 2009, a monkey fitted with a BrainGate was able to operate a robotic arm with sufficient dexterity to turn a water valve each time it required a drink.

Warwick believes humans will soon be using brain-linked cybernetic limbs, and he doesn't think it's going to stop with medicine. He foresees a time when cybernetic implants will also provide a range of optional modifications to the body. For a start, linking the human brain to computers will vastly improve its ability at mathematics, enabling it to not only solve horrendous equations but, for example, to visualize what a ten-dimensional universe might look like from the inside, a feat way beyond the capabilities of our present, organic brains. He envisages brains linked seamlessly to the Internet, giving them rapid access to almost limitless amounts of information. He even says that our brains could be linked directly to one another, enabling them to communicate faster and more effectively by transmitting thoughts as electronic data rather than the inefficient audible signals we use when we speak. And that's to say nothing of the physical advantages in, say, strength and athletic speed that cybernetic components may ultimately bring.

Warwick believes that there will be strong motivation for us to modify our bodies in these ways, as a consequence of developments in artificial intelligence (see Chapter 23). As machines become smarter than us—not only able to outperform us in simple number-crunching tasks but also able to out-think us—we will, only naturally, seek ways in which we can keep up. And if we can't do that organically, it stands to reason that we will consider plugging into ourselves the same technology that we're trying to compete with. If you can't beat them, join them.

Those who don't will risk becoming second-class citizens, says Warwick. What we are likely to see is a gradually widening racial divide between those with cybernetic enhancements and those without. As the "withs" become smarter, thanks to all their bolted-on technology, they're going to become ever-less interested in what the "withouts" have to tell them. When was the last time you had a good discussion with your cat about politics, quantum physics, or football? Similarly, is an advanced cybernetic organism—accustomed to communicating with its colleagues at high speed via electronic data transfer—really going to take much notice when a human comes up to it and starts making these strange, primitive noises that we call speech?

There's a strong parallel with the way we treat animals. "We tend to treat other intelligent creatures—like chimpanzees or cows—pretty awfully, really," says Warwick. "I think the same would be true here. I can quite see future cyborgs, and the Cybermen of *Doctor Who*, not particularly wanting to be friends with humans."

Even the methods that the cyborg ruling class use to reproduce and raise their offspring will be better and more efficient than the way humans do it. Cyborg children are likely to receive their first implants at an early age. Putting an implant into the brain of a newborn baby would be controversial in today's climate. But in terms of maximizing the implant's benefit, sooner is no doubt better than later. Warwick draws a parallel with the way in which children today have an almost natural aptitude with computers, because they are exposed to them from a very early age, compared with those of us who first used computers much later in our lives.

While acknowledging the ethical concerns, Warwick thinks that there will come a time when cybernetic body parts will be implanted into babies while they are still in the womb. "If a baby is born with different sensory inputs and different mental abilities, it will learn to use those abilities better than if it picked them up later on," he says. The children of the cyber elite could, literally, be born cyborgs.

In *Doctor Who*, the only time we see the creation of a Cyberman in any detail is during the Sixth Doctor adventure "Attack of the Cybermen," when the mercenary character Lytton gets captured and partially converted. But it seems likely that these ruthless cyborgs would have no ethical qualms whatsoever over modifying their young from the earliest possible age—particularly if that meant the results would be better.

Golden Bullets

So we've seen how technology can be, and is being, added to organisms. We've seen how that technology can make the organisms stronger and more efficient. And we've seen how it could eventually turn the resulting cyborgs against their purely organic forebears.

The question is, if you're unfortunate enough to be one of those purely organic forebears, just what exactly can you do about it? Luckily for us, the Cybermen do have several weaknesses—radiation, solvents, and the precious metal gold.

Radiation is well known to have a deleterious effect on electronics. In fact, this is the basis of so-called electromagnetic pulse (EMP) weapons. An EMP can generate high voltages (sometimes thousands of volts) in electrical conductors and equipment, damaging or even destroying sensitive components. Nuclear explosions give off considerable EMPs. Indeed, the biggest nuclear weapons are designed for this purpose, and are capable of knocking out electronic systems—particularly communications—over continent-scale areas.

Other kinds of radiation consist of high-energy subatomic particles. These can cause problems in digital electronics, such as those found inside a computer. A high-energy particle smacking into, say, a computer memory chip can flip the state of one or more of the electrons being used to store data, resulting in a software glitch or even a hardware error. Beyond the protection of our planet's atmosphere and magnetic field, interplanetary spacecraft are pummeled by particle radiation. That's why most of them are clad with shielding—to guard their electronic systems against it. Such shields, however, would be too bulky for a Cyberman to lug around (at least without becoming a sitting duck for the nearest UNIT soldier to pick off).

Organic solvents are another weapon. Thrown over them, these melt the Cybermen's plastic components. It's quite true—organic solvents like acetone, turpentine, and ethanol do have a nasty effect on plastics. (Pour some nail polish remover into a Styrofoam cup if you don't believe me—though don't try this anywhere near your best carpet.) The Cybermen's weakness to solvents was exploited to great effect in the Second Doctor serial "The Moonbase."

If those two fail, there's a final, albeit somewhat pricey solution—not silver bullets, but gold ones. The *Doctor Who* writers have made gold toxic to the Cybermen, claiming that it clogs their respiratory systems. Humans in the *Who* universe took advantage of this weakness during the "Cyber Wars" (when, in the twenty-sixth century, many planets united to eliminate the Cyberman threat) with the invention of the glittergun, a weapon that fires gold dust at its targets.

This is a tougher one to explain in terms of real science, though. While some heavy metals are quite considerably toxic—notably beryllium, lead, and mercury—gold generally isn't. In fact, it's not absorbed by living tissue, making it relatively harmless (why else would it be used so extensively in dentistry?). It's a very good conductor (which, coupled to its excellent corrosion resistance, is why it's often used in music speaker connectors). So you could perhaps argue that it gets into the Cybermen's electronics somehow and short-circuits them. The trouble with that idea is that all your frontline cyber-soldier would need to protect himself is a roll of insulating tape so he can make sure he has no bare wires exposed.

The most promising explanation seems to be the bad skin reaction that some people experience to gold jewelry. Sufferers find that contact with the metal brings their skin out in a rash. It's possible some kind of similar reaction takes place when gold dust gets into a Cyberman's lungs. If you could unfold and lay out the porous interior of a human lung, it would be vast—an area of about 70 square meters. That's a huge surface for the dust to react with. If the tissue of these fine pockets and cavities became swollen it could restrict airflow and cause suffocation. But even if the Cybermen haven't thought of wearing an artificial respirator—like a diver's aqualung, say—so they don't have to breathe the gold dust in, there's still a problem with this. The human "gold allergy" isn't actually a reaction to gold at all. It's a reaction to the small amounts of nickel used to bulk up lower-purity gold. I say "small amounts"—9 karat gold is actually 63 percent nickel by weight. And reactions to nickel affect roughly 16 percent of all people.

But there is a . . . well, a lining to this particular cloud. If your gold jewelry is irritating you, then you probably need to go out and treat yourself to some new stuff—18 karat or more ought to do it.

But don't throw away all those 9 karat rings and necklaces just yet. You never know—all that nickel might just come in handy when the Cybermen invade.

10

The Daleks

"What am I dealing with? Little green men?"
"No—little green blobs in bonded polycarbide armor."

—Group Captain Gilmore and the Seventh Doctor,
"Remembrance of the Daleks"

A pepperpot brandishing a plunger isn't how most people would have imagined the ultimate force for evil in the Universe. The Daleks debuted in the show's second-ever adventure. Called "The Daleks" and starring William Hartnell as the Doctor, it was broadcast between December 1963 and February 1964. Since then this menacing cyborg race has become synonymous with the *Doctor Who* series and has even acquired its own entry in the *Oxford English Dictionary*.

The Daleks were created by writer Terry Nation and BBC designer Raymond Cusick. Nation claimed that his inspiration came from watching the Georgian State Dancers—the women had long skirts that reached the floor, making them appear to glide as they moved across the stage. Nation and Cusick have to be credited for their bravery. This was less than 20 years after the end of the Second World War, and beaming images into people's homes of an authoritarian master race seeking to conquer the Universe to the cry of "exterminate!" could so easily have ended in disaster. Nation freely admitted that he had based his creation strongly on the Nazis.

So what are the Daleks? The misconception that many *Who* virgins have is that the Daleks are robots. They're actually cyborgs—the genetically modified descendants of the Kaled race, embedded within mechanical bodies. The Kaleds and Thals had been fighting a long and protracted war on the planet Skaro, using nuclear and biochemical weapons. These weapons had caused genetic mutations in the Kaleds, which their chief scientist, Davros, accelerated and honed to create a race that was devoid of compassion or remorse, but fiercely obedient and ruthlessly focused on its objectives—the perfect soldiers. Davros encased each of his genetically modified master beings inside a tanklike "travel machine." This was a human-sized armored vehicle with an eye, some sort of a "gunstalk," and an arm so that each Dalek could manipulate its environment. The arm usually has a suction cup on the end, but claws, trays and even flamethrowers have appeared in the past.

More about the travel machine later. First of all, how do you go about genetically engineering the perfect life-form?

Being Davros

The public's perception of genetic modification didn't get off to the best of starts. When supermarkets tried to introduce fruit and vegetables that had had their genes tinkered with in order to lengthen shelf life, to make them more resistant to bugs and frost, or even to improve their nutritional value, most shoppers reacted with revulsion. Fears that renegade genes might somehow jump from the food to the person eating it, or that modifying an organism in this way could lead to toxicity or other unforeseen effects, saw genetically modified, or GM, produce whisked off the shelves almost as swiftly as it had arrived.

GM Daleks aren't all that different from GM potatoes. At least, the procedures to make them would be very similar. One way to do it would be to use a virus. These are tiny bundles of

DNA and protein that work by hijacking a cell's ability to replicate. When you catch a common cold, the cold virus latches onto cells in your body and injects its genes in among your own. The cell's copying machinery is then fooled into building new copies of the virus from the genetic blueprint that's been injected. These new viruses burst forth and go on to infect other cells.

Biologists have found that they're able to manipulate viruses, smuggling extra genes into them so that when the virus implants its genetic material into a host cell, these extra genes get inserted as well. So the genetic make-up of the cell has been modified. This method is already being researched as a way to treat defective-gene disorders in humans. But it's not without its dangers. In September 1999, a clinical trial of the technique ended in tragedy when a gene therapy virus inserted its new genes into the wrong place in a patient's DNA, resulting in his death.

The other problem with this procedure is that it affects only cells that the virus can reach. This is fine if you want to use genetic modification to cure, say, a lung disorder. Delivering a virus to the lungs is simple enough—ask anyone who's had the flu. But what about a brain disease? "Getting it into the brain is hard," says Denis Murphy, a biotechnologist at the University of Glamorgan in Wales. "So for treating something like Alzheimer's you'd have to think of a different way."

This is a big problem for the Daleks, because unless you can get the virus into reproductive cells—sperm or egg cells—then the genetic modifications that it causes won't be passed on to future generations. So what biotechnologists often do—rather than genetically modifying adult animals—is to modify the creatures while they're still just egg cells. Applying the virus to an egg cell in culture leads to an embryo in which every cell is modified. Implanting this embryo back into the female animal that the egg was originally taken from, and then bringing it to term, creates a new baby animal that has the new DNA in every cell of its body and that will also pass that DNA on to its offspring.

That's one way of doing it. Yet the *Doctor Who* writers seem to be suggesting that it was done another way. Davros took Kaled mutants that had been created in the war with the Thals and selectively bred them into Daleks. So it could be more realistic that the Daleks were created by a genetic selection process rather than active genetic modification.

It's rather like the process that's discussed today for creating "designer babies." This is where a couple use in vitro fertilization to produce a number of embryos in a test tube. These are then screened, and only those that are free from any disease-causing genetic mutations are implanted into the mother and brought to full term. The technique enables parents to screen offspring for heritable diseases such as Huntington's and cystic fibrosis, as well as genetic conditions like Down syndrome.

In the same way, Davros could sift through the genetic mutations caused by the radioactive and chemical fallout from the war and select those giving the qualities he wanted in his super-soldiers. "What you're doing is increasing the probability of getting an intelligent child from say 1-in-100 to 1-in-3," according to Murphy. "The Daleks would probably then grow up the three kids and only one would be really intelligent, so they'd get rid of the other two."

This is essentially the same process that evolution enforces on us naturally. Random factors introduce mild mutations into our genes and only those mutations leading to traits that make us better able to survive get passed on. Genetic selection is just an accelerated form of that process, and it's very similar to the way the *Doctor Who* writers had Davros create the Daleks. In theory, scientists could use the technique to select the right genes needed to create perfect soldiers, super-fast athletes, or even blue-eyed blondes.

But be careful what you wish for. Davros engineered the Daleks to be utterly devoid of emotions such as compassion, empathy, pity, and remorse. Yet while removing these qualities might possibly be useful in the heat of battle, would you really want your species to be without them 24/7? As we saw in Chapter 1, optimal behavioral strategies are a delicate balance between selfishness and cooperation.

"The Daleks were a society," says Murphy. "And as such they would need to work together and trust each other." Even in battle, there's a place for compassion (e.g., dealing with the injured) and camaraderie (e.g., looking out for your comrades). These qualities didn't emerge as nature's way of being nice. A soldier may think he's showing kindness when he helps an injured comrade, but that's just nature's way of making sure he's not standing alone come the next battle. Similarly, he covers the backs of his fellow soldiers in battle in the hope that they will do the same for him. The emotions that drive us to these actions have evolved because they confer a survival advantage (indirect as it may be)—and genetically selecting them out might not be such a smart idea after all.

But let's assume Davros got it right and succeeded in creating a race that really is superior, which he then unleashes upon the galaxy. What should be the advice for the inhabitants of any planet getting in the way of his invading Dalek army? "Get out of the way!" might seem like the first instinct. But just because the Daleks have superior abilities doesn't guarantee them victory. "You sometimes get more intelligent organisms that are outcompeted by rats and bacteria," states Murphy. It sometimes comes down to numbers, but it can often be a consequence of the "superior" species fighting in an alien environment that it's not adapted to—exactly the situation that would face an invader from another world.

So a wily Davros might do well to make his Daleks "generalists," able to perform well in a wide range of conditions and environments. Humans are a prime example of a generalist lifeform. We're good with technology yet we can also live without it; we can live anywhere from tropical jungles to frozen polar wastes; we can eat meat or we can eat plants. "Cows eat grass and lions eat antelope—they can't do vice versa. But we can do almost anything," says Murphy. This degree of versatility isn't found anywhere else in the animal kingdom and must explain to a large degree humans' rise to dominance on Earth.

It's clear what we have to do then: stop fretting about the Daleks' invasion of Earth and get plotting the human invasion of Skaro. And the first thing we're going to need are weapons that can penetrate a Dalek's armor. So what exactly do we have to get through?

Bullet Time

One of the best bits added to the Daleks for the 2005 relaunch of *Doctor Who* was their power to dissolve bullets in midair. The effect was rendered in glorious "bullet time" by London-based postproduction house, The Mill.

Something very similar has already been developed by scientists working for the UK Ministry of Defence. So-called electric armor can protect tanks and other armored vehicles from rocket-propelled grenades (RPGs).

RPGs are formidable weapons. The warhead packs a shaped explosive charge, which on impact squirts a jet of molten copper into its target at speeds of around 1,000 mph, capable of punching a hole through over 30 centimeters (about a foot) of conventional steel armor.

The new electric armor consists of two layers of metal armor separated by a layer of insulation. The inner metal layer is charged to a high voltage; the outer layer has the thickness of conventional tank armor and is earthed. If an RPG penetrates the outer layer and breaks the insulation, a connection is made and thousands of amps of electricity flow through the molten copper, vaporizing it. In tests, an armored personnel carrier fitted with electric armor survived repeated RPG attacks that should have destroyed it several times over.

Electric armor aside, the Dalek's main body casing is tough enough to deflect bullets and blast fragments on its own. Modern tanks and armored personnel carriers use composite armor with layers of ceramic to blunt armor-piercing projectiles and softer steel to absorb the energy of high-speed impactors. Some types of armor even have spaced layers, with hollow cavities inside, that offer protection against shaped-charge jets. These jets dissipate naturally after a short distance, so increasing their travel distance through the armor lessens their impact.

In the Seventh Doctor adventure "Remembrance of the Daleks," the Doctor states that Dalek armor is made of "bonded polycarbide." No such thing exists in the real world, but the "poly-" prefix suggests that this is some kind of polymer—a plastic. It's certainly true that as polymer science improves, so lightweight plastics and composites are increasingly being used in roles where in the past only metal would do.

A current example is in the aircraft industry, where lightweight alternatives to metals are being used everywhere from jet fighters to airliners. In the news at the end of 2005 was the Airbus A380 superjumbo, a behemoth capable of lifting over 550 passengers into the sky at a time—largely because of the lightweight materials that it's built from.

The U.S. military is already experimenting with polymer-based vehicle armor. They're investigating a rubbery coating (made from polyurethane and polyurea) that's sprayed over conventional steel armor to spread the force of an explosion across the structure underneath. In tests conducted by scientists at the U.S. Office of Naval Research, a 230-kilogram (500-pound) bomb detonated next to two trailers simply dented the one that was sprayed with the polymer, yet completely destroyed the one that was not.

Another neat trick from modern tank design that the Daleks have pinched is to use a steeply angled "glacis" plate—the thick armor plate immediately in front of the turret. Angling this plate makes bullets and other projectiles more likely to glance off. It also makes the armor effectively thicker. Imagine a bullet traveling horizontally through a sheet of steel. If the steel is vertical, then a horizontal line from one side of the plate to the other is as short as it can be. But now tip the plate so that

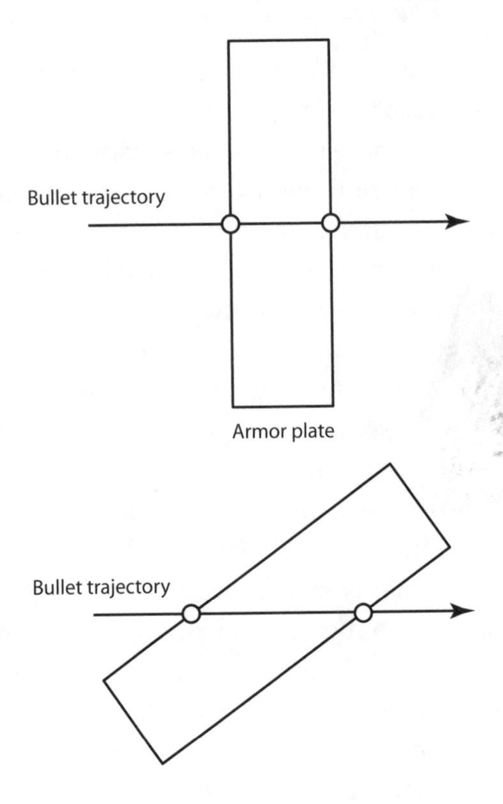

FIGURE 8. The front glacis plate of a battle tank's armor is steeply angled, not just to deflect incoming projectiles but also because tilting the plate makes the armor effectively thicker. Perhaps this is why the Daleks also have steeply angled armor plates protecting their "travel machines."

the bullet's horizontal trajectory effectively makes a diagonal through it, and the line gets a lot longer (figure 8).

Evil Eye

So we've got our work cut out for us. With all that armor and force fields to get through, making a hole in a Dalek is no easy task—as Captain Jack Harkness's small band of human resistance fighters discover in "The Parting of the Ways." But Daleks do have a weakness—shoot them in the eye stalk.

The eye stalk is an electronic sensor that feeds visual information to the Dalek mutant inside the travel machine. Research has demonstrated that electronic cameras really can feed images directly into the brain's visual cortex. In February 2005, Claude Veraart, at the Catholic University of Louvain in Belgium, announced that he had implanted prototype bionic eyes into two blind patients. The devices work by recording images just as a video camera does and then relaying them to a coil wrapped around the patient's optic nerve, where they are converted into electrical stimuli. In 2009, a blind man was able to see light for the first time in 30 years after being fitted with a bionic eye at London's Moorfield eye hospital.

It's fascinating research. Yet the really amazing possibility with electronic eyes is only just starting to be realized. And this is that they could enable a person (or a Dalek) to see not just in visible light, but at other wavelengths as well—such as infrared, ultraviolet, or maybe even high-energy radiation such as x-rays or gamma-rays.

Imagine being able to sense the world in ways that other creatures cannot. Murphy points out that scientists already use electron microscopes to look at tiny creatures at a magnification of as much as a million times. And soldiers now routinely use night-vision goggles that detect infrared radiation. These devices could be miniaturized to fit in a mechanical "eye" just like the Dalek eye stalk. The ability to "see" in these wavelengths would enable the user to see through walls and deep inside the bodies of other creatures. Virtually nothing would be secret any more.

In 2002, Kevin Warwick, the cyberneticist at the University of Reading who we encountered in Chapter 9, experimented with ultrasound vision. As we've seen, he had a radio receiver-transmitter chip implanted and surgically linked to the nerve fibers in his left arm. The chip was used to link a host of computers and electronic equipment with his brain. Then he took signals from an ultrasonic sensor and fed them through the chip and into his nervous system. Although the results were crude, he found that he could effectively "see" in ultrasound.

"It took about 6 weeks for me to learn to recognize the current pulses we were sending in," says Warwick. "After that it was just like having another sense and my brain was quite happy with it. I had a blindfold on and I was wandering around the lab quite happily."

At one point, one of Warwick's colleagues feigned an attack on the professor. Using nothing but his new ultrasound vision, Warwick was able to detect and react to the threat, ducking almost instinctively.

But our humble brains have evolved with just five senses. Would they be able to cope with all the extra data from an array of sensors relaying sound and wavelengths of light from beyond the visual spectrum? Warwick thinks so. He suggests the brain would spread its processing power intelligently across the senses, depending which senses were most important at a given time. In the same way that you might not hear someone talk to you when you're engrossed in a good book, so perhaps smell or taste might temporarily diminish to let us make the most of a night-time vista in infrared.

You would think the Daleks would use their genetically enhanced brainpower to "see" in every waveband of sound and light available. And yet in the Second Doctor serial "The Power of the Daleks," the Doctor and two other characters, Janley and Valmar, were able to fool a Dalek into believing that the three of them were dead—an act that elementary infrared vision would be able to see through straight away. You also have to wonder whether Davros bothered to include an active radar or sonar on his Daleks—otherwise gauging distance with just the one eye stalk could be difficult.

Let's give him the benefit of the doubt (again) and assume that he has. In that case, what do you do when a Dalek has you in its sights and is trundling after you? Hit the stairs, of course.

Elevate!

Everyone knows the best way to give a Dalek the slip is to head for higher ground. Perhaps the biggest irony of the Daleks' success (they were such a ratings winner that at one point producers ruled that every *Doctor Who* season must feature them) was how impractical their travel machines really were. All you needed to halt the ultimate power for evil in the Universe was a flight of stairs, some small obstacles, or even just a bit of bumpy ground. Clearly, unless their conquest was to be limited to the Universe's bungalows and ranch houses, something had to be done.

And it was. By the 1989 Seventh Doctor story "Remembrance of the Daleks," the Doctor's wheeled adversaries had obtained the means to fly—an ability that was reprised in the Ninth Doctor episodes "Dalek" and "The Parting of the Ways."

But just how might this work? Some kind of jump jet—similar to the engine used in a plane capable of short vertical takeoffs and landings—would be one option, though this would require a considerable amount of fuel. Another possibility is antigravity—a way to literally block out the gravitational pull of the Earth or whatever planet the Dalek was standing on.

In 1996, Eugene Podkletnov, a researcher at Tampere University in Finland, announced that he had built the world's first antigravity machine.

Podkletnov claimed that his 275-millimeter-wide superconducting disk, made from yttrium-barium-copper-oxide, when cooled to around −230°C and spun up to 5,000 rpm, reduced the weight of objects placed above it by up to 2 percent. He was also careful to stress that he had prevented air currents and magnetic effects from influencing the experiment.

The trouble is that no one has yet managed to replicate Podkletnov's work. That's not for lack of trying. NASA, BAE Systems, and Boeing, along with teams from a number of universities, have all attempted to verify the findings—without success. Podkletnov claims that this is because the experiment is very difficult. Yet he's being extremely cagey about the details, withdrawing the only scientific paper explaining how to do it just weeks before it was due to be published.

One researcher did claim to have predicted Podkletnov's result as early as 1989. Ning Li, then at the University of Alabama, devised a theory explaining how superconductor systems can absorb large amounts of energy—enough to produce measurable changes in the strength of gravity. However, until her theory can be demonstrated experimentally—and Podkletnov's research independently confirmed—it looks as though the stairs will remain the best way to thwart a Dalek.

11

The Slitheen

"Excuse me, do you mind not farting while I'm saving the world?"

—The Ninth Doctor to "Joseph Green," "Aliens of London"

Flatulent aliens with a taste for human flesh and an evil plot to sell off the Earth as radioactive fuel made for a good yarn in the consecutive Ninth Doctor adventures "Aliens of London" and "World War Three." The hostile aliens were the Slitheen—a family of bloated, oversize creatures from the planet Raxacoricofallapatorius.

The Slitheen planned to take over the Earth by killing off and then impersonating high-ranking politicians, such as Britain's acting prime minister Joseph Green. To disguise themselves, they used human body suits. But how do you fit an 8-foot-tall, incredibly fat alien into a human skin? To do this, the Slitheen all wear "compression field" generators around their necks, which shrink their bulky bodies down to the size of a chubby human, enabling them to squeeze into their disguises.

Miniaturization technology like this has featured in the show several times before. Back in the classic series, the weapon of choice of the Doctor's evil counterpart the Master was his "tissue compression eliminator"—which slays its victims by shrinking them. In the Fourth Doctor story "The Invisible Enemy," microscopic clones of the Doctor and his assistant Leela explore the inside of the real Doctor's body. And in "Carnival of Monsters," the "miniscope" was a futuristic peepshow containing miniaturized life-forms from across the galaxy. So how do you go about shrinking stuff?

Shrink Fit

The first strategy the Slitheen might consider is physically squashing their bodies down to smaller size. The trouble here is that bones, tissue, and blood aren't all that compressible, because the interatomic and intermolecular forces inside them are enormous. Go find a large pair of pliers and try compressing your thumb—it'll start to hurt long before any noticeable reduction in volume has taken place.

Alternatively, because the atoms and molecules in your body are locked together by electromagnetic forces, another strategy might be to use electric or magnetic fields to condense them down into a smaller volume. But you'd have to have a darn peculiar kind of field to do this. The problem here is that the Slitheen would have to somehow squeeze down the spacings between their atoms and molecules without squeezing down the actual atoms and molecules themselves (as this would affect their function). This, too, is bit of a nonstarter.

If they can't cram their atoms and molecules together more tightly, then what about getting rid of a few of them? Again, they're going to run into trouble with this one. The atoms and molecules in our bodies are all there to perform specific functions. True, some of us may have a few more cells in certain places than we might like, but in general all of our bits are there for a reason. In particular, if you wanted to shrink yourself by removing material, then you would have to remove some of the atoms that make up your DNA molecules. But DNA stores all of your genetic information—nature's blueprint for how the body works. Whittling away at this could have a marked effect on the essence of the person.

A slightly more ambitious scheme still might be to try to reduce the size of atoms by dinking around with the laws of quantum physics. The size of an atom is fixed by a number called the "fine structure constant." Known to physicists by the Greek letter alpha (α), this number fixes the strength of the electromagnetic force. Its value—the oddly precise one one-hundred-thirty-seventh—is inversely proportional to the size of atoms, so one way to make atoms smaller is to make alpha bigger.

If α is meant to be a "constant" of nature, then how can it change? In 1998, a group of astronomers from universities in the UK and Australia came forward claiming to have uncovered evidence for just that. They had studied quasars, fiercely bright galaxies at the edge of the observable universe. During the course of its journey to Earth, some of the light from a quasar gets absorbed by intervening clouds of gas. The absorption takes place at particular wavelengths, or equivalently, colors of the light. So when astronomers split the quasar light up into its spectrum—the band of colors similar to what you see when there's a rainbow in the sky—the absorption shows up as dark lines at specific, well-defined colors in the spectrum. The positions of the dark lines are fixed by two things: the atomic structure of the particular type of gas (e.g., hydrogen, helium, etc.)—and the value of α. Alpha controls the separation between pairs of dark lines—the bigger it is, the farther apart they are. The astronomers looked at quasar light absorbed by clouds at different distances from Earth, finding that the separation of the lines decreased the further away the cloud was. Because the light-travel time between the cloud and the Earth increases with distance, that seemed to imply that α was smaller in the past.

Recently, however, measurements by some of the world's most powerful telescopes have cast doubt on this idea. In particular, studies with the European Southern Observatory's Very

Large Telescope (VLT) in Paranal, Chile, have now placed very tight limits on how much α can have varied by, constraining it to less than 0.6 parts per million over the last 10,000 million years. "Astronomers have looked and there is no variation of alpha with time," insists Graham Thompson, a professor of physics at Queen Mary, University of London.

Could we make α change ourselves? There is one way you could do it. When the Universe was born in the heat of the Big Bang, it was in a highly "symmetric" state. There was no gravity or electromagnetism—all the forces of nature were unified into a single almighty superforce. But as the Universe cooled, the forces that we recognize today gradually broke away in a series of events called "phase transitions." During a phase transition, the strength of the force that's breaking off is determined by a semi-random quantum process known as "spontaneous symmetry breaking." Physicists have likened this to a donkey standing equidistant between two rows of carrots. This is the donkey in its symmetric state. However, unless the donkey is to starve to death, it must break the symmetry and eat from one row or the other. Which row it chooses is effectively determined at random. Similarly, in spontaneous symmetry breaking, the Universe randomly chooses values for the constants of nature from the range of values that are permitted by quantum physics.

So one way to change α could be to restore the electromagnetic symmetry and then break it again. The only trouble with doing this is that it would mean heating up the object you want to shrink to the temperature of the Universe during the electromagnetic phase transition. This happened one ten-billionth of a second after the Big Bang, when the temperature was a sultry million billion degrees C. At that kind of temperature, most objects would be instantly vaporized into their constituent subatomic particles—the fundamental building blocks of all matter.

For this reason, Graham Thompson thinks it's unlikely that we'll ever be able to fiddle with the fine structure constant, least of all inside living matter (assuming we want it to stay living).

"You'd have to create an enormous energy density locally, thus recreating, almost, Big Bang conditions, and then tip the difference so alpha turns out larger," he says. "I'd be delighted to have a seminar from the Master on how he does this at will."

A million billion degrees could certainly explain the lethality of the Master's favorite weapon. But just to make doubly sure no one gets out alive, raising the value of α will also completely disrupt the biochemistry required for life. For example, doubling α increases the energy required to break electron bonds—a key process in chemistry—by a factor of four, and this will prevent life-sustaining chemical reactions, which govern everything from the digestion of food to the movement of muscles.

All this doesn't do much to help the poor Slitheen squeeze into their human body suits. But there's even worse news in store for them. The *Doctor Who* writers have given the Slitheen chronic wind—it's not natural flatulence, but a side effect of their compression fields. As matter is squashed down, the writers reasoned, gas gets squeezed out. Oddly, the gas smells like bad breath. It's not a sign that the Slitheen are talking from the wrong end, but rather a hint at what they're made from. Whereas life on Earth is based on the chemical element carbon, the Slitheen are meant to have evolved from a calcium-based biology. That's why their farts are supposed to smell like rotting teeth.

But I smell something else here.

Calcium Chemistry

Carbon turns out to be a very good molecule for life for a number of reasons. Firstly, it has a property that biologists call *tetravalence.* This means that it has four electrons available for *covalent* bonding with other atoms and molecules.

A covalent bond is a strong but flexible chemical bond created by atoms that share some of their electrons with one another. Remember those stick-and-ball molecule models at school? Tetravalence simply means that carbon is a ball with four sticks coming out of it. This means that it's possible to build tens of thousands of complex three-dimensional molecules from it, with a rich range of properties.

Calcium, however, is only bivalent—it has just two electrons available for making covalent bonds. Bivalent bonds can form only simple line-like molecules, and so no complex molecular scaffoldings are possible. Calcium can form *ionic* bonds instead, where atoms become bonded together by the attraction of opposite electrical charges. But these are very much weaker. "It would be impossible, therefore, for calcium to form semi-rigid open structures like protein or DNA," says Matthew Genge, an astrobiologist at Imperial College London.

And even if calcium-based life could somehow get started, the odds of it surviving are slim. Every now and again, Earth is battered by a cataclysm, a global natural disaster such as an asteroid impact or a giant volcanic eruption. Life here survives only because the richness of carbon chemistry gives it the machinery it needs to evolve and adapt to the new environments that are left behind after these catastrophic events.

"Thanks to evolution you can have a system which is more complex, so you create biodiversity," says André Brack of the Centre for Molecular Biophysics in Orléans, France. "Then when you have a cataclysm, the chances to survive are greater."

It's unlikely then that the highly basic, uniform sort of biology arising from calcium could sustain life over a sufficient number of generations for it to evolve from microbes into larger animals. And that means that while the Slitheen make great TV (especially when they explode on contact with vinegar), for the time being that's where they're staying.

12

The Autons

"Think of it. Plastic—all over the world—every artificial thing, waiting to come alive. Shop window dummies, phones, the wires, cables . . ."
"The breast implants . . ."

—The Ninth Doctor and Rose, "Rose"

The Autons, living plastic mannequins, clashed with both the Third Doctor, in the serials "Spearhead from Space" and "Terror of the Autons," and the Ninth Doctor, in "Rose." The Autons are controlled by the Nestene Consciousness, a collective intelligence that fell to Earth during a meteor shower and that has the ability to animate plastic and turn it into living material.

According to scientists, breathing life into plastic isn't all that difficult. Researchers at the University of Sheffield, in England, are studying plastics that can change shape in a precisely controlled way. They've been using these materials to construct artificial muscles that contract in response to a control signal, turning chemical or electrical energy into mechanical force.

Plastics consist of long, thin molecules known as polymers, which exist in the form of tangled spaghetti-like networks. When dissolved in a solvent, a polymer can do one of two things—it can either coil up and shrink or stretch out and expand. Chemists are able to control this reaction by changing the solvent's pH (its acidity) or by applying an electric field. If the change causes the polymer to expand, then it can exert a force that pushes; whereas if it causes the polymer to contract, then it makes a force that pulls—just like a real muscle.

"There are many ways of inducing shape change in a polymer," says Tony Ryan of the University of Sheffield's Polymer Centre. "Some really clever polymers will change their shape in response to a local change in the concentration of specific ions present, like calcium. Other polymers can have special links put in them so that they change shape when exposed to light."

The artificial muscles created in this way are still quite feeble, offering about 1 percent of the efficiency of real biological muscles. "There was a recent competition for a synthetic muscle which could compete in arm wrestling," says Norman Billingham, a polymer chemist at the University of Sussex. "From what I can remember, a 14-year-old girl easily beat all entrants."

But Ryan believes that it will be only a matter of time until synthetic muscles are as powerful as the real thing and that one day it will be possible to build a human-sized plastic robot with power requirements no greater than those of a real person, just like an Auton.

He imagines that robots built from this plastic could be controlled by what he calls microfluidics, in which microscopic channels are used to pump acidic fluid around the body, to change the pH levels in specific regions and so control the robot's muscles. "Or you could just rig it up to wires to change the pH locally with electricity," he says.

Building a large robot purely out of Ryan's plastic isn't possible. The robot needs the additional support of a rigid plastic or metal skeleton. It's no accident that biology employs much the same strategy, using small motors, which we call muscles, to move levers, known as bones. "The same combination of soft and wet muscles with hard rigid bones should prove to be optimal for an Auton," says Ryan.

Such robots could even power themselves in the same way we

do, using the energy-transfer molecule ATP (adenosine triphosphate). Research in 2005 led by Ryo Yoshida at the University of Tokyo has shown how enzymes in plastic can generate energy from ATP, in turn causing the plastic to change its shape. Yoshida's team supplemented synthetic plastic with a natural enzyme to drive the reaction, but in future they say the whole system could be synthetic. They found ATP to be the most efficient way to store energy for these synthetic chemical motors, and future research may well prove that it can't be bettered.

This close parallel with real biology could be no accident. Many components of our bodies are already polymer based. For instance, the DNA molecules that store our genetic information, the proteins that repair our cells, and the peptides that hold the proteins in our bodies together are all examples of natural "biopolymers," perfected over tens of thousands of years of evolution. So for the best example of living plastic, perhaps all we have to do is take a look at ourselves.

Mind Control

The Autons were animated by the Nestene Consciousness, a being that fell to Earth inside a shower of hollow plastic meteorites. It brought the Autons to life using a technique called "telepathic projection," which sounds rather like what parapsychologists refer to as psychokinesis, or PK—the ability claimed by some to be able to move objects around purely by the power of the mind. Could there really be something in such claims, or are they just nonsense?

Parapsychologists make a divide between what they call macro-PK and micro-PK. Macro-PK is the big stuff, like levitating tables and bending spoons, while micro-PK describes the phenomenon applied on much smaller scales—such as using PK to influence dice-rolling experiments.

"The problem with all the macro-PK claims is that they can all be duplicated by fraudulent techniques—and often are," says Chris French, head of the Anomalistic Psychology Research Unit at Goldsmith's College, University of London. And the scientific tests that have been conducted so far have failed to produce positive results.

"Micro-PK, on the other hand, is much harder to fix," says French. "But because the effect is much tinier, it's also much harder to tell if anything's actually going on—you need to do lots of trials and then apply statistical analysis to them to see if an effect is taking place."

Some micro-PK experiments do seem to show positive results. Intriguing research led by Robert Jahn of the Princeton Engineering Anomalies Research group at Princeton University has found that test subjects seem able to use PK to skew the output of random number generators. These machines use either electronic noise or radioactive decay to produce what's meant to be a stream of completely random numbers. Jahn's group, however, found that the output isn't quite as random as you would expect. The effect is tiny, amounting to a deviation of just one fiftieth of a percent from random chance. But Jahn claims that this is down to the psychokinetic influence of his test subjects.

At the moment there seems to be no way other than PK to explain this, leaving skeptics somewhat baffled. "It's intriguing, it's controversial, and it's not widely accepted by mainstream scientists," says French. "Essentially we just don't know what to make of it. Watch this space."

So the evidence seems to be consistent with some degree of micro-PK but not in favor of macro-PK. Sadly, making plastic mannequins walk around is very much macro-PK. So it looks like bad news for the Nestene Consciousness.

Of course, different species have different mental abilities. And it may be that there are creatures on a planet somewhere that have new and unusual neurobiologies, giving them mind powers over and above those of mere human beings. But without any kind of evidence to back it up, this is pure guesswork—not science.

If Mother Nature can't deliver the goods, could technology

possibly lend a helping hand? Already, quadriplegic patients have been given brain implants allowing them to control their computers using only their thoughts (see Chapter 9). What if such an implant were connected to a radio-frequency transmitter, so that brain impulses could be beamed over a distance? If these impulses could then be picked up by a radio receiver connected to items of machinery, then this would be a kind of PK. It could almost certainly be used to control robots such as the Autons.

In the far future, nanotechnology would perhaps be the way to implement such technology non-invasively. It's been suggested that nanorobots—machines measuring on the scale of 100 to 1,000 one-billionth of a meter across—could enter the bloodstream and swim to the brain, where they could then do the same job as a larger cybernetic implant.

Robert Freitas, Senior Research Fellow at the Institute for Molecular Manufacturing in Palo Alto, California, says that injecting 10 billion nanorobots—one to monitor each of the 10 billion neurons in the human brain involved in data processing—would add just 200 milligrams to your gray matter's total weight and that the tiny machines would generate just 2 watts of waste heat. By comparison, the body can lose heat at a rate of up to 100 watts when it's in a cold room and still around 10 watts in a relatively warm environment.

Freitas suggests that a fleet of nanorobots in the brain could empower us with other kinds of "psychic" abilities as well—not just PK but also mind reading, precognition, and even dowsing. If that's true, then never mind the Nestene Consciousness. Robots on the brain could be technology's way of letting us all be Uri Geller.

13

The Silurians and the Sea Devils

"This is our planet. We were here before man. We ruled this world millions of years ago."

—Silurian leader, "The Silurians"

Dinosaurs running amok have done the rounds in science fiction on plenty of occasions, so technologically advanced humanoid dinosaurs packing energy weapons and trying to take over the world certainly makes for a refreshing change. In "The Silurians," Jon Pertwee's Third Doctor meets the Silurian race—reptilian humanoids that went into hibernation millions of years ago and have now awoken to reclaim the Earth as rightfully theirs.

The Silurians were so named because they were meant to be from the Silurian period of Earth's prehistory, about 440 million years ago. But this just couldn't happen. The trouble is that 440 million years ago is way before any form of higher life was thought to exist on the Earth's landmasses—fossils from this period indicate that the most advanced terrestrial life-forms were the primitive ancestors of spiders and centipedes.

But it gets better. Realizing this too late to change the script for "The Silurians," the show's writers included some dialogue to accompany the creatures' next appearance (in "The Sea Devils"), in which the Doctor explains that they should more accurately be called Eocenes, as that's the epoch that they're really from. But this doesn't work either! The Eocene period was between 34 and 56 million years ago, but the dinosaurs disappeared from the Earth—probably following the impact of a large asteroid or comet (or hijacked space freighter; see Chapter 18)—65 million years ago, and so were actually long gone by the time of the Eocene.

In the Third Doctor adventure "The Sea Devils," and later in the Fifth Doctor adventure "Warriors of the Deep," the Silurians are joined by their aquatic cousins the Sea Devils. Also scientifically advanced reptilian humanoids, these amphibians hailed from bases on the ocean floor.

Both races had voluntarily placed themselves into hibernation when a passing planetoid threatened to disrupt the Earth's atmosphere, rendering the planet temporarily uninhabitable.

Deep Sleep

Many animals on Earth hibernate. It's a state of self-induced hypothermia in which a creature's metabolic rate, body temperature, and breathing are all greatly lowered to enable it to conserve energy and so survive without eating during times when food is scarce—such as in the winter. Some species have been known to hibernate for long periods. "Brine shrimp eggs have gone for 100,000 years," says biologist Jack Cohen. "And the evidence is pretty good for a lot of bacteria and some nematode worms going for five or six million years. But not more."

If the Silurians really are meant to be from the Eocene epoch, then they must have hibernated for at least 34 million years—and, for reasons outlined above, perhaps longer. Could artificially induced hibernation be the way to do this? In 2005, *New Scientist* magazine reported that researchers at the Fred Hutchinson Cancer Research Center in Seattle had managed to reduce the metabolic rates of mice by as much as 90 percent, placing them in an artificially induced state of hibernation.

When the team, led by Mark Roth, exposed the mice to hydrogen sulfide (the same gas that makes farts smell) they found that the animals' body temperature fell from 37°C to just 15°C and that their breathing dropped from 150 breaths per minute to just two. Roth's team used a mixture of normal air with hydrogen sulfide mixed in at a concentration of 80 parts per million. The mice were kept in suspended animation for 6 hours, after which time the air supply was returned to normal and they awoke unharmed.

How can this be explained? Philip Bickler, an anesthetist at the University of California, San Francisco, thinks that the hydrogen sulfide is fooling the hypothalamus region of the brain—the area that controls temperature—into cooling the mice down, which in turn slows their metabolism. Roth disagrees. He thinks the gas is interfering with the body's production of the energy molecule ATP. Making ATP requires oxygen, and so if the mice produce less ATP, that reduces oxygen demand, lowering their breathing and therefore their metabolic rate.

"We are, in essence, temporarily converting mice from warm-blooded to cold-blooded creatures, which is exactly the same thing that happens naturally when mammals hibernate," said Roth. However, he and his team declined to comment on the maximum length of time that hibernation could ultimately be sustained using the technique.

The Silurians and the Sea Devils had planned to wake from their hibernation once the Earth's atmosphere had returned to normal after the passage of the planetoid. But in "The Silurians," the Doctor realizes that the disaster never actually happened—and that the planetoid became the Earth's Moon instead.

Moonrise

This sounds quite similar to the Big Splash theory, which is regarded by astronomers as the best theory for the Moon's origin. It says that a body roughly the size of the planet Mars crashed into the Earth a little over 4.5 billion years ago, shortly after our planet had formed. The planetoid struck the Earth a glancing blow, ejecting into space a cloud of pulverized material—consisting of the planetoid itself and some of the Earth's mantle. Some of this material was captured by Earth's gravity, forming a ring of debris around the planet that condensed to form the Moon in around 10 years.

The theory is supported by studies of Moon rocks brought back by the Apollo astronauts. These show the presence of material very similar in composition to the Earth's mantle, and that the rock has a surprisingly low content of volatile material—suggesting that it's been strongly heated. Additionally, measurements of the Moon's overall density and studies of its seismological activity show that its metal core is much smaller than would normally be expected, consistent with the idea that much of it is formed from the metal-poor mantles of the Earth and the planetoid.

Clearly though, there are a number of discrepancies between the elements of this theory and the events described in "The Silurians." Firstly, the Big Splash theory dates the arrival of the lunar planetoid at more than 4.5 billion years ago. Compare that with the date the Silurians could have gone into hibernation—somewhere between 34 and 440 *million* years ago.

And then there's the fact that the fictional planetoid never actually hit the Earth—it was simply captured by the planet's gravity. This doesn't square with the Big Splash theory at all, although perhaps that's just as well for Earth in the *Who* universe. If something the size of Mars had crashed into the planet that recently—even 440 million years ago, as far back as we can push it—then it's unlikely there would be any intelligent life on Earth today. The surface would have been completely melted and reformed. Evolution's reset button would have been well and truly pressed, and we'd be lucky if even the most basic forms of life had started to reappear on Earth by the present era.

Whether the Silurians had chosen to hibernate or not, they most definitely would not be here to reclaim the planet as their own.

14

The Sontarans

"If we cannot control the power of the Time Lords, then we shall destroy it."

—Sontaran Commander Stor, "The Invasion of Time"

A vile, warlike race of squat, bald aliens, the Sontarans have crossed paths with the Doctor on five occasions so far: "The Time Warrior" (1974), "The Sontaran Experiment" (1975), "The Invasion of Time" (1978), "The Two Doctors" (1985), and the two-part Tenth Doctor adventure "The Sontaran Stratagem" and "The Poison Sky" (2008).

The Sontaran species come from a massive, dense world with a strong gravitational field, which explains their compact, stocky form—if their bodies were too slight in such an environment their bones would break whenever they fell over. The Sontarans are also unusual because they are asexual. They don't reproduce sexually like humans and other mammals do on Earth—that is, by mating with a partner of the opposite sex. Sontarans have only one gender, and their children have only one parent.

There are life-forms that do this on Earth, such as freshwater hydra, which give birth to their young as "buds" that break away from their bodies and then grow into full-sized creatures. But it's not just invertebrates. Some bigger animals are asexual too, such as the North American whiptail lizard. And there are insects, for example, stick insects, that are parthenogenic—only females of the species are needed to reproduce.

In many ways, it's more efficient to reproduce asexually than sexually. It's quicker and uses less energy and the chances of damaging complications arising are smaller. In fact, it's so advantageous that the reason why so many creatures chose to reproduce sexually instead is hotly debated.

"Ask modern biologists why we have sex, and you'll get five different answers," says biologist Jack Cohen at the University of Warwick in England.

One reason is that it improves a species' chances of survival when conditions on its planet become suddenly inhospitable. Climate-changing events known as cataclysms—planet-wide natural disasters such as giant asteroid impacts and super-volcanic eruptions—punctuate the fossil record of the Earth (and presumably of other worlds too). If creatures are to survive in the aftermath of a cataclysm, they must be able to evolve and adapt to the new environment, which demands that they be able to rapidly change their traits and abilities, and hence their DNA (which carries our genetic information—everything from the color of our eyes to how many arms and legs we have—from generation to generation).

Sexual reproduction is one way to achieve the genetic diversity needed to recover from a cataclysm. Offspring created sexually have a mix of DNA from both parents, allowing large variations to be introduced from one generation to the next. Through random chance, some of these variations will inevitably confer advantages that help individuals of the species survive a cataclysm. Offspring created asexually, on the other hand, have identical DNA to their parent—they are exact copies, and so don't enjoy the same advantage when disaster strikes. Studies also show that sexual species are better able to fight disease, compared with their asexual equivalents. The idea that sex carries an evolutionary advantage is sometimes known as the Red Queen hypothesis, inspired by a character in Lewis Carroll's

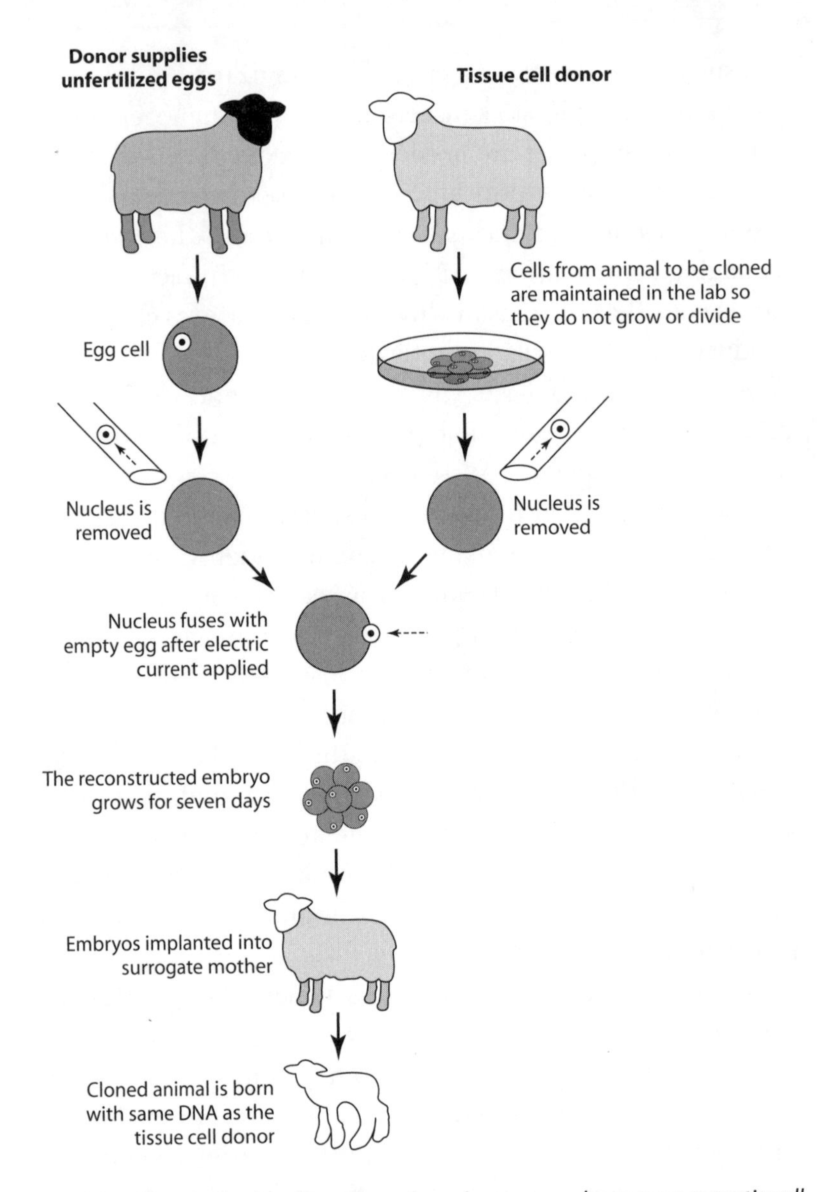

FIGURE 9. How Dolly the sheep was cloned, a process known as *somatic cell nuclear transfer*. The genetic material is removed from an egg and replaced with material from the animal to be cloned. The egg is then stimulated to make the cells start to divide, and the resulting embryos are implanted back into the mother animal that the egg was taken from.

Through the Looking Glass. The Red Queen takes Alice to a place where the landscape is moving rapidly, and it's necessary to run fast to keep up with it. This has become a metaphor for the evolutionary "arms race" between a creature and its environment. It's been argued that the substantial reshuffling of genes from generation to generation in sexual reproduction is what enables creatures to stay one step ahead in this race.

It's perhaps no surprise, then, that on Earth we see asexual reproduction only in lower life-forms; maybe all creatures that do this are wiped out every time there's a cataclysm or the outbreak of a serious infectious disease, and so never get the chance to evolve into higher, more intelligent beings.

The Sontarans could get around this problem by taking control of their reproductive process. According to the *Doctor Who* writers, they don't undergo natural asexual reproduction—they clone themselves. It could be that as part of the cloning process each generation has its DNA checked for any weaknesses that may have emerged and then modified if need be—in just the same way that biotechnologists today might genetically modify a potato to make it frost-resistant.

In "The Two Doctors," we learn that the Sontarans can mass-clone themselves at rates of up to a million embryos every four minutes. The clones then take just ten minutes to grow to adulthood, so the Sontarans don't have any "children" to speak of—or, at least, not for long.

We're also told that the first successful cloning experiments took place in the year 3922. In reality, cloning has happened on a much faster timescale than that. The first cloned mammal, Dolly the sheep (see figure 9), was born at Edinburgh's Roslin Institute in 1996. Dolly was created using a process called *somatic cell nuclear transfer.* First an unfertilized egg cell is taken from the mother animal. The cell's nucleus—the bit containing all the genetic information—is then removed and replaced with the nucleus of a cell from the animal to be cloned. (In this case, the sheep was cloned from a mammary cell, so the scientists cheekily named her Dolly, after Dolly Parton.) The complete egg is then stimulated electrically to make its cells start to divide. The resulting embryos are grown in a Petri dish for 5 to 6 days, and those that appear to be developing normally are finally implanted back into the mother, where they are brought to full term and delivered as normal.

This is how Dolly was born. She enjoyed approximately 2 years of normal life—but then the problems began. Studies by the Roslin team in 1999 suggested that she was ageing prematurely. They found that Dolly's telomeres—caps on the end of each DNA molecule, rather like the plastic on the end of a shoelace that stops it unraveling—were shorter than they should be for an animal of her age. Telomeres are strongly implicated in the aging process, progressively shortening as a cell grows older, until the telomere is so short that the cell can no longer divide and may die. The theory was that because Dolly had been cloned from a 6-year-old ewe, she had been born with telomeres that were already this age and so already substantially shortened. In terms of the wear and tear on her cells, Dolly was 6 years old when she was born.

More evidence has since emerged to back up this idea. Sheep normally live to an age of 11 to 12 years; however, Dolly died in February 2003 at the "age" of just 6. She was killed by a lung infection—a common ailment in older sheep. And a year before she died she was diagnosed with arthritis—again, normally found only in older animals. Yet, oddly, premature aging seems not to be a problem with many other animals that have been cloned.

Cloned beings in the *Doctor Who* universe don't enjoy long lives either. We're told in "The Invisible Enemy" that this is due to "psychic stress problems." While the proposed cause is questionable, the symptoms were undeniably borne out in studies of Dolly.

Dolly's birth was followed up in 1998 by the creation of some 50 cloned mice by a team in Hawaii. Various animals have been cloned around the world since, including horses and dogs. There has also been talk of using cloning to resurrect extinct animals such as mammoths and the Tasmanian tiger, *Jurassic Park*-style, by implanting samples of their DNA into the egg cells of closely related species still alive today. Scientists, however, are divided over the feasibility of such plans.

More important is the question: if we can do it with animals, then can we do it with people too? Human cloning is a thorny issue. Italian embryologist Severino Antinori has long made clear his intention to perfect human cloning to help infertile couples have children. This work remains highly controversial and is unlikely to be endorsed by the international scientific community any time soon—at least not until the issue of premature aging, and a raft of other medical and ethical concerns, have been explored and addressed fully.

One form of human cloning, however, has been given the go-ahead. "Therapeutic" human cloning was given the green light by the UK government in August 2004. This is a process in which cloned human embryos are harvested for *embryonic stem cells*—cells which are yet to differentiate themselves into particular tissue types such as brain, blood, and liver.

It might sound unpleasant, but the embryos used are only a few days old—literally a tiny ball of cells. Stem cell treatments offer new hope to sufferers of degenerative illnesses such as Parkinson's. Doctors coax the stem cells to grow into new brain tissue, reversing the damage done by the disease. The embryos are cloned by somatic cell nuclear transfer, using genetic material taken from the patient. This makes the resulting stem cells an exact genetic match to the patient and so ensures that there's no chance of them being rejected by his or her immune system.

Of course, you might wonder exactly what use somatic cell nuclear transfer will be to the Sontarans, with no females and so no female reproductive components to cut and paste genetic material into. It seems likely, though, that a space-faring race as advanced as they are would be able to construct artificial female organs, either cybernetically or more likely through biological engineering.

Whether the Sontarans get around to it or not, humans are already on the case. In vitro fertilization—the fertilization of human eggs surgically extracted from their mother's body—has been going on since 1977. And in 2002, researchers at Cornell University's Center for Reproductive Medicine and Infertility in Ithaca, New York, announced that they had built an artificial human womb for embryos to develop in. In experiments, embryos were able to attach themselves to the walls of the surrogate womb, which the doctors had built from real womb lining cells. The cells were cultured on a specially shaped scaffold, which dissolved once the tissue had grown around it. Hormones and nutrients were then added to the cell structure to create an environment extremely close to a real human womb. Scott Gelfand, an ethicist at Oklahoma State University, told the UK's *Observer* newspaper at the time: "Some feminists even say artificial wombs mean men could eliminate women from the planet and still perpetuate our species."

Sontaran Snacks

Asexual reproductive cloning isn't the only weirdness of Sontaran society. Rather than eating food like any normal person, a Sontaran warrior feeds by taking in energy through a port on the back of its neck, known as the "probic vent." The probic vent is one of the species' great weaknesses—they can be stunned by a blow to it or even killed if stabbed in it. But could it ever really be used to eat through?

The problem with this is that we obtain more than just energy from our food. Even if a species were suited to taking in calories as raw energy (rather than the fat and carbohydrate that we get our energy from), it would still need to find a source of protein (from which we build new cells to replace damaged ones) and a source of essential vitamins and minerals—the chemicals that we need for healthy living.

True, plants consume raw energy—photosynthesizing to "feed" on sunlight. But they supplement this process by taking in carbon dioxide from the atmosphere and nutrients from the soil around them.

Biophysicist André Brack thinks that all life-forms would need to take in matter as well as energy. He likens the Sontaran feeding process to a lamp fed by a solar panel and a battery. "During the night, the battery will feed the lamp, *but this is not alive*," he says.

Other scientists aren't so sure. Norman Maclean, a geneticist at the University of Southampton in England, points out that life around deep thermal vents in the ocean floor feeds purely off the chemicals spewing from the vent, with little regard for conventional nutrients. "I don't agree that all life-forms require matter," he says.

Even if the Sontarans were able to subsist purely on their "energy charge," they would still need to have a very bizarre internal biology, unlike anything we've seen so far. "Perhaps they would replace the intestinal system with the organic equivalent of batteries," says Stuart Clark, author of *Life on Other Worlds and How to Find It*.

Brack thinks that food might be the least of the Sontarans' worries, however. He's of the opinion that the biggest threat to the survival of this warlike race will come, not in the form of malnourishment, but through its own self-destruction. History is replete with sorry tales of aggressive, greedy civilizations that picked one fight too many.

So the Sontarans might do well to note, as they chow down on their evening energy charge—he who lives by the disruptor rifle will very probably also die by it.

15

Martians, Go Home!

"Well, what do you feel about all this, Walters? Bet you didn't think you'd have ice monsters and things like that to deal with when you volunteered for the job, did you . . . ? Well, did you?"

—Britannicus Base Leader Clent, "The Ice Warriors"

Life-forms from Mars laying siege to the Earth have been a science fiction staple ever since H.G. Wells first unleashed his savage, unearthly heat ray on the unwary town of Woking in the 1898 novel, *The War of the Worlds*.

Never ones to miss a trick, the writers of *Doctor Who* have come up with their own Martian invaders, the Ice Warriors, which debuted in the Second Doctor story of the same name in 1967. Led by Varga, the armored reptilian humanoids were discovered frozen in their spacecraft in a glacier on Earth during the next ice age, circa 3000 AD, by a team from Britannicus Base—a human outpost in the icy wilderness.

The Ice Warriors are giants, towering over the human occupants of the ice station. This is broadly accurate. Any large animal living on Mars would be taller than similar life-forms on Earth. Back here, our legs and bodies need to be short and thick to support our weight in Earth's gravity. But due to its smaller mass, the gravity on Mars is only one third what it is on Earth. And so the same objects there weigh just a third what they do here. With less weight to support, creatures on Mars would be likely to grow taller and spindlier.

But that's about where the good news ends. The Doctor states that the Ice Warriors breathe nitrogen, because it's the gas that most of Mars's atmosphere is made up of. But that's simply not true. Mars's atmosphere is actually over 95 percent carbon dioxide, or CO_2. That means that no kind of terrestrial animal life could breathe there.

It's also rather chilly on Mars. True, we are discussing "ice" warriors here, but nighttime temperatures on Mars can fall as low as –100°C. The coldest temperatures that large creatures on Earth can withstand for long periods are in the Antarctic seas, where fish produce "antifreeze proteins" to lower their freezing point. Yet even here the temperature dips only to a relatively balmy –6°C.

So nothing will be able to breathe, and anything that can will be frozen solid at night. As if that weren't bad enough, there's no ozone layer on Mars—so the planet is blasted by harmful ultraviolet radiation from the Sun. And to make matters worse, there's only a tiny magnetic field—a feeble effort about one ten-thousandth the strength of Earth's. Our planet's magnetic field fends off high-energy cosmic rays (harmful radiation from space).

It's possible that the red planet was more hospitable than this in the past. Scientists speak of a "warm, wet era" in Mars's history. But current thinking is that this was very short-lived, much shorter than the billions of years that it took for macroscopic life to emerge on Earth.

The Red Seed

The Ice Warriors returned in the 1969 story "The Seeds of Death," in which they brought seed pods from their planet to Earth. Inside the pods was a deadly fungus that saps oxygen from the atmosphere—their plan being to make Earth more like Mars.

Could Martian plant life really exist? On Earth, plants require soil that's rich in nitrogen. To date, the nitrogen content of Mars's soil hasn't been measured—although this gap in our knowledge will be addressed by the forthcoming generation of sample return missions. These will gather Martian soil samples and fly them back to Earth for a full laboratory analysis.

Plants would suffer in the harsh environment on Mars, in exactly the same way as animal life. That said, mosses and lichens do thrive at high-altitude locations on Earth, where the thin atmosphere lets through a lot of UV radiation. These plants grow on the underside of rocks, where they are shielded from the radiation. Maybe the same could happen on Mars.

Of course, all this assumes that we are talking about life "as we know it." But this cannot be the case as far as the Ice Warriors are concerned. They want their plants to make Earth more like Mars, which in the real world means converting oxygen to CO_2, but photosynthesis in terrestrial plants does exactly the opposite. It takes CO_2, water, and sunlight and converts it into sugars plus oxygen. This process also casts doubt on whether there could be any significant quantities of plant life on Mars today—if there were, then where's all the oxygen that should be in Mars's atmosphere as a result?

So the case for plants and animals on Mars isn't looking good. But let's give them the benefit of the doubt one last time. Let's suppose that there is some radically different kind of biology that leads to life-forms that thrive in the conditions on Mars. If that were the case and there were macroscopic life-forms prowling the surface of the red planet—silicon-based rock monsters, or whatever—then where exactly are they? Orbiting space-probes have now imaged Mars down to a resolution of less than a meter. If the Martians are there, they clearly don't get out much.

Although things may look bleak for plants and animals on Mars, there's one form of life that we may yet find there: microbes. In 1996, scientists claimed to have found evidence for fossilized bacteria in the meteorite ALH84001, a lump of rock that fell to Earth from Mars 13,000 years ago. Nowadays, most researchers are highly skeptical of this claim. But other evidence means that the case for Martian bugs is still a strong one.

Firstly, scientists can now say—with 100 percent certainty—that liquid water once existed on Mars. Readings taken by NASA's twin Mars Exploration Rovers have revealed layered salt deposits consistent with those laid down during the slow evaporation of saline water. They've also found "spherules," spherical particles formed by minerals once dissolved in groundwater but that later settled in the cavities of porous rock.

But the absolutely clinching evidence came in 2008, when NASA's *Phoenix* spacecraft landed on the planet. *Phoenix* used its robot arm to dig a shallow trench in the Martian soil. The trench was just a few centimeters deep, but that was enough. It revealed small white chunks buried in the soil and was able to watch as they evaporated away in the Martian sun. When *Phoenix* carried out detailed tests on these white chunks it found that they were indeed made of water ice.

The second piece of evidence for Martian microbes comes from the European Space Agency's Mars Express probe, which has found methane gas in the planet's atmosphere. Methane is a known by-product of biological activity. The quantity is tiny: about ten parts per million. But it's significant because methane is known to break down rapidly, combining with water to make carbon dioxide. The fact that Mars Express has found any at all means there must be a methane source on Mars that's keeping the atmosphere topped up with the gas.

The detection has since been confirmed by astronomers using the European Southern Observatory's Very Large Telescope, in Chile. And Mars Express has now gone farther, finding evidence that the methane sources seem to be confined to particular geographical areas on Mars, rather than being evenly spread around the planet. Nevertheless, the finding remains controversial—some scientists argue that the true culprit may be global warming on Mars causing natural underground deposits of methane ice to melt and evaporate.

Jim Bell, a Mars expert at Cornell University, thinks that if microbes are the source of the methane, then they must be hiding even deeper underground. Here, in the warmth of the planet's interior, safely screened from the radiation at the surface, life could thrive. In fact, underground is where most of the biomass is located on Earth. Bell believes the next step is to send lander missions that will dig deep into Mars's surface to search for these bugs.

This will be a major undertaking. "It's going to require human crews bringing drilling equipment like you would use at a field site on Earth," he says. "They'll dig down tens of meters, hundreds of meters to where it starts getting warm and where there may be liquid water."

Ironically enough, then, the truth about whether there really is life on Mars must wait for the Earthlings to invade. Current estimates put this somewhere around the year 2030.

Pyramids of Mars

No discussion of the Doctor's adventures on Mars would be complete without mentioning one of the most highly regarded *Doctor Who* adventures of all time: "Pyramids of Mars." Airing in 1975, the serial told the tale of the villain Sutekh (also known as the Egyptian god Set) who was imprisoned under a terrestrial pyramid 7,000 years ago by Horus and the Osirians, using a force field generated by a pyramid on Mars. What the program-makers didn't realize was that less than a year later NASA would discover a real pyramid on Mars. Images returned in July 1976 by NASA's *Viking I* orbiter showed pyramid-like structures in the Cydonia Mensae region of Mars's northern hemisphere. Along with the pyramids there was a group of smaller objects that was soon named the "city" and a mound with features resembling a human face—the now-infamous "face on Mars." Conspiracy theorists seized upon the discovery, hailing it as evidence for alien activity on Mars—possibly linked to the ancient Egyptian civilization on Earth.

All such thinking was scotched, however, when NASA's Mars Global Surveyor spacecraft flew over the site again in 1998. Its higher-resolution cameras revealed the face and the pyramids for what they really are: natural rock formations. It's a cautionary tale—proof, if ever it was needed, of just how easy it is for the human eye to be fooled.

16

The Krynoid

"The sergeant's no longer with us. He's in the garden. He's part of the garden."

—Mad botanist Harrison Chase, "The Seeds of Doom"

It's estimated that the first land plants emerged on planet Earth about 500 million years ago. It's often said that plants are the lungs of the planet, taking in carbon dioxide and giving off oxygen that animal life-forms such as human beings need to breathe. But so often in science fiction, *Doctor Who* included, our onetime leafy allies have ended up turning against us.

Take the evil Krynoid, for instance. This was a "galactic weed" that the Doctor and Sarah Jane encountered in "The Seeds of Doom." Its spores fly through space, looking for host planets. When they find one, the spores land and grow into Krynoid plants, which devour the planet's own plant life and convert all the animals into more Krynoids. In "Terror of the Vervoids," the Sixth Doctor tangles with bipedal vegetable life-forms that have even learned to talk. They dislike humans and attempt to kill them at every opportunity by firing poisonous thorns at them.

There has been one welcome let-up in this onslaught of hostile foliage—in the Ninth Doctor episode "The End of the World." Here, in the far future, the Doctor meets the trees from the Forest of Cheem, a benevolent humanoid species who are the descendants of Earth's rainforests.

Which is the more likely scenario—hostile plants or friendly ones? I don't know about you, but I want a couple of questions answered before I'm prepared to even consider a plant's motives or demeanor. One: how can they be intelligent? And two: even if they are, what the heck are they doing walking around?

The Fuchsia's So Bright

Never mind 5 billion years from now. Some botanists are coming to the conclusion that plants today are already showing the signs of primitive intelligence. In an interview in March 2005, Leslie Sieburth of the University of Utah told the *Christian Science Monitor*: "If intelligence is the capacity to acquire and apply knowledge, then, absolutely, plants are intelligent." In May of that year, biologists from around the world gathered in Florence for the first-ever conference on "plant neurobiology"—a new scientific discipline to investigate the similarities being discovered between the physiology and anatomy of plants and features of animal nervous systems.

"Plants have an immune system, they have hormones just like we do [chemical messengers that regulate their growth], they can communicate with each other, and they can manipulate animal behavior in some clever ways," says biotechnologist Denis Murphy, of the University of Glamorgan, in Wales.

For instance, some plants appear to mimic insects in order to encourage pollination. Acacia trees go further, producing unpalatable tannin to stop animals from grazing on them. The scent from grazed leaves can then be picked up by other acacia trees, causing them to start producing their own tannin long before the animals have even arrived.

Perhaps the most demonstrably impressive example of plant intelligence is the Venus flytrap. This species is especially clever, not simply because it can catch flies, but because of the way it does it. Each trap will spring only when one of the tiny hairs that act as triggers inside the trap is touched twice in a short interval. This is a safeguard just in case the hair is knocked accidentally, say by a falling raindrop (because each trap can open and close only a limited number of times before it withers and dies). To do this, the plant must be able to remember which hairs have been recently triggered—it has a primitive memory.

Some biologists are now finding that plants have similar chemical pathways to those that are present in animal nervous systems—and that allow animals to process, store, and transmit information. A family of twenty genes has been found in plants that instruct them to manufacture what are known as ionotropic glutamate receptors (iGluRs). In animals, these proteins sit on the outer membranes of nerve cells where they play a central role in the transmission of nerve impulses. Plant neurobiology is such a new field that biologists don't yet know for sure whether these receptors take the same role in plants. However, work is under way to find out.

Research at the University of Lancaster in England has shown that the roots of plants are very sensitive to low concentrations of glutamate, which is the same molecule that animals use to transmit signals in their central nervous systems. The Lancaster team believe that plants use it as a signal to tell them where to find nutrient-rich patches of soil—all decaying animal and vegetable matter is rich in glutamate. This, they think, causes the roots to begin a kind of foraging behavior, with increased root branching in the vicinity of the glutamate source.

"There is one interesting parallel with us humans who are also very sensitive to glutamate," says plant biologist Brian Forde, one of the Lancaster researchers. "Our tongues are about as sensitive to glutamate as some plant roots are. For us, this is how we recognize protein-rich foods. So it's possible to think of roots 'tasting' glutamate in the soil and responding accordingly."

If plants do have nervous systems, then could that mean they have brains as well? Back in 1880, Charles Darwin suggested that plant roots could be harboring a kind of primitive brain. But this now seems unlikely. "Plants don't have the equivalent of a brain that coordinates the activities of the whole organism," says Forde, because they don't need one. Animals use their brains and central nervous systems to rapidly coordinate the movements of the body's extremities. But plants don't depend on such fast, coordinated movement for their survival.

Could they still be conscious without a brain? Again, unlikely. We don't even assign consciousness to some life-forms that do have a brain, such as earthworms. Indeed, consciousness isn't even necessary for many complex behaviors. Most of the decisions that we make are taken without thinking. We breathe and blink automatically. And anyone who drives knows how easy it is to switch off and let instinct take over.

Plants could evolve brains in the future, though. Conditions on Earth are ever changing, and it's not impossible that there could one day become an evolutionary benefit to plants developing brains and eventually conscious thought. "If dinosaurs became birds, then who knows where trees and other plants could go," says Peter Barlow, a biologist at the University of Bristol in England.

Of course, there's little use having an intelligent plant that's capable of making decisions if it can't actually act on those decisions and move around. But this takes energy, and there's only so much a plant can extract from photosynthesis—the process of converting carbon dioxide gas and sunlight into food. "We could have people engineered to have green skin who would not need to eat so much—the problem is that we animals use so much energy that it would be difficult to generate enough photosynthesis just from our skin," says Denis Murphy.

Seth Shostak of the Search for Extraterrestrial Intelligence (SETI) Institute in Mountain View, California, and Margaret Turnbull of the Carnegie Institution of Washington, D.C., have highlighted just what a woefully small amount of energy photosynthesis actually generates. The sunlight reaching a square meter of the Earth's surface provides about 100 watts of power. However, only about 35 watts of that is actually absorbed by the plant. The process of photosynthesis itself is so inefficient that only a quarter of that, about 9 watts, actually gets turned into plant food. Let's assume we have a human-sized plant that isn't too fussy about wearing clothes—so that its total surface area exposed to the Sun is about 3 square meters. It'll be generating about 27 watts of power. Now compare that with the energy requirements of a moving animal of the same size as a human being. Humans require roughly 2,000 calories a day, which, converting to slightly more scientific units, equates to about 100 watts for every hour in the day. Taking into account that photosynthesis can work only when the Sun's out suggests that a mobile human-sized plant would require around ten times as much energy as it's capable of generating.

That said, there are animals on Earth that do photosynthesize, albeit indirectly. The convoluta worm only ever eats once in its lifetime—but when it does, it feeds on special algae and is subsequently sustained by starches made photosynthetically by the algae under its skin. In return, the algae get a safe haven in which to live and grow—it's a symbiotic relationship.

Fade to Green

The case of the convoluta worm lends some weight to another leafy theme that the *Doctor Who* writers have seized on. "It shows plant-animal hybrids aren't just science fiction," says Denis Murphy. Human-plant hybrids featured in the Fourth Doctor serial "The Seeds of Doom," in which the Krynoid seed pods were able to gradually mutate human victims into plants.

Even in labs, such creations are not a new idea. As long ago as 1976, a team of researchers at Duke University in North Carolina succeeded in fusing cultured human cells with stripped-down cells from a tobacco plant. The human cells retained their integrity for up to six days after the fusion.

It's food for thought: is there any limit to the range of species that cellular material can be transferred between in this way? Perhaps in the far future, humans will benefit from plant-like modifications.

"One scenario might be a post-apocalypse world, where all the crops have been wiped out along with many edible animals," says Murphy. "Food would be very scarce, so the ability to generate our own food would become extremely useful." In this scenario, he envisages humans merging with algae which would deliver sugars and starches into the bloodstream photosynthetically, in much the same way as the algae in the convoluta worm do. We would possibly become shorter to conserve energy and would be prone to developing mutations that maximize skin surface area, such as growing extra fingers with "leafy" webs of skin between them.

"In time the hybrids may even become completely sessile [immobile] and resemble bushlike plants," he says. "These bush folk might still be able to speak and could wistfully recall the days when their ancestors walked the Earth and built cities."

For the moment, though, plants are starting to look rather different from the motionless, passive life-forms we're used to, sitting on the window ledge sucking up water and sunshine. They're emerging as dynamic and sensitive organisms that are highly active and capable of competing for limited resources.

So maybe it's time you spoke nicely to your house plants a little more often—or at least remembered to water them now and again. One good turn usually deserves another, and they might not be rooted to the spot forever.

17

Stupid Apes

"Homo sapiens. *What an inventive, invincible species. It's only a few million years since they crawled up out of the mud and learned to walk. Puny, defenseless bipeds. They've survived flood, famine, and plague. They've survived cosmic wars and holocausts. And now, here they are, out among the stars, waiting to begin a new life. Ready to outsit eternity. They're indomitable.*"

—The Fourth Doctor, "The Ark in Space"

Modern humans are the species known as *Homo sapiens*—Latin for, ironically enough, "wise man." The prevailing thought is that we emerged from the older species *Homo erectus* around 200,000 years ago in Africa, from where we then spread around the world.

Humans are a relatively recent addition to the planet, which is why the human skull unearthed in 12-million-year-old volcanic sediment in the Fourth Doctor adventure "Image of the Fendahl" really is eons out of its time (as the series correctly stated). Back when this ancient ash layer was being deposited, humans weren't even a twinkle in the eye of ancient primates. In fact, it would be another 7 million years before the blood line leading to modern humans was believed to have broken away from that of modern chimps.

The earliest actual humans were members of the species *Homo habilis,* which emerged about 2 million years ago, in the Paleolithic era. The Doctor and his companions visited this time in "An Unearthly Child," the first episode of the show ever broadcast.

Humans have evolved staggeringly since the Paleolithic era. Our bodies have grown more upright and more athletic and, most important of all, our brains have grown much larger. We have emerged from apes into a highly advanced, intelligent species. Many people have wondered what we might evolve into next. Some have even asked whether we will continue to evolve at all. In the past, people with certain genetic weaknesses have been more prone to die before they could have any children to pass their weak genes on to. That way the weakness is eliminated from the gene pool and the species evolves to become stronger. This is called natural selection. Nowadays, more advanced medical technology and geriatric care have reduced the lethality of these genetic weaknesses, so the "weak" genes remain in the gene pool.

Tough as the old days were, our species was stronger for it. But as technology increasingly enables us to dodge the cruel hand of natural selection, does this mean we will cease to evolve?

On the contrary. Evolutionary biologist Christopher Wills of the University of California, San Diego, says there's growing evidence that human evolution is actually accelerating. "It is true that people who might otherwise have died can now survive, but such relaxation of selection will have little effect on our gene pools over hundreds or even thousands of years," he says. Instead, Wills thinks that evolution is taking off apace in other areas. The genes that are changing the fastest are those involved in brain function. As our application of technology makes our environment increasingly complex, so our mental abilities are improving in order to deal with that.

Wills adds that as well as overall brain power, the evolution of our species is also selecting for diversity in brain function. "While none of us can do everything well, all of us can do something well," he says. Diversity in brain function keeps the human race stocked with specialists in every field, from nuclear physicists to musicians. He thinks that this trend will continue in the future.

Technology may also be used to modify humans directly. As we saw in Chapters 9 and 10, to some extent genetic and cybernetic modifications are already empowering the human species to steer the course of its own evolution. Some researchers, notably technology guru Ray Kurzweil, have even suggested that technology could soon allow us to totally transcend our reliance on our biological bodies. Wills advises caution with robotic and genetic modifications, however. "My guess is that the law of unintended consequences will continue to dog our attempts to change our genetic make-up for a long way into the future," he warns.

All things considered, then, what are humans going to look like in the future? Wills doesn't think we'll notice much of a physical change in the form of our species over the next few thousand years. But over millions and (assuming our species survives that long) billions of years, the changes could be amazing—especially if humans branched out to colonize other planets, where different environments will place different evolutionary pressures on our genes.

This is the story that seems to be borne out in *Doctor Who*. In the near future, humans don't really seem to have changed all that much—at least, not according to what we see of the Doctor's travels in that era. But push the boat out a little farther and skip ahead a few billion years and things are very different. In "The End of the World," the Ninth Doctor and Rose Tyler travel forward in time 5 billion years to watch the Sun swell up at the end of its life and devour the Earth. There, on the space station *Platform One,* they meet Lady Cassandra O'Brien Dot Delta Seventeen: the last human being.

Actually, she's not quite the last human. She considers herself to be the last remaining "pure" human—all other members of the species have interbred with life-forms from other planets.

Given the difficulties with interspecies breeding on Earth, such interplanetary cross-matching seems rather improbable. Wills believes it's more likely instead that we will try to make other animals more like us by using genetic modification techniques to insert our genes into their DNA. He wonders if such biologically engineered beings might replace electronic and mechanical assistants as the computers and robots of the future.

But back to Cassandra. She has had 708 pieces of cosmetic surgery carried out on her body and as a result now resembles just a piece of skin with eyes, a nose, and a mouth, stretched taut across a metal frame. She's wheeled around on a cart by two attendants, her brain sitting underneath in a nutrient tank. Cassandra's thin skin is prone to drying out and as a result she has to be moisturized with a spray gun almost constantly by one of her attendants.

Perhaps reassuringly, Wills thinks this is an improbable scenario. "I don't think that Lady Cassandra is likely," he says. "After 5 billion years, we should surely be able to live without being moisturized all the time."

Hiding behind the Sofa

Humans are often motivated by fear. Whether it's fear of failing an exam, fear of being harmed, or just fear of walking home through the churchyard after dark making us take the longer route home instead, fear dictates many of our decisions. Perhaps our unique perception of fear is all part of being human. Other creatures experience fear, but humans seem to have taken the emotion and lifted it to a whole new level of complexity. Sure, any sane mouse will flee when it sees a cat, but how many mice do you know that fret about social anxiety? Perhaps it's one of the vulnerabilities that endear us to the Doctor—he makes no secret of the fact that humans are his favorite species in the Universe.

Fear and suspense have always been elements of the show. Indeed, taking refuge behind the furniture has been a core part of the *Doctor Who* experience for younger viewers of all generations. But the "sofa cliché"—for those few fans who don't know about it, this is the all-too-human habit of hiding behind the sofa to avoid looking at something scary but having to peek at it anyway—seems most common among those who grew up with the Fourth Doctor, played by Tom Baker. Storylines then were arguably at their closest to the horror genre and the show was at its farthest from mainstream children's entertainment. Scenes of violent death were common, monsters were frequent and horrific, and even the title sequence had something slightly chilling about it.

It was also during this era that Mary Whitehouse's National Viewers' and Listeners' Association took issue with the program, in particular one scene from "The Deadly Assassin" that showed the Doctor being drowned and that has consequently been cut from the BBC's master copy of that story. More recently, the BBC has come under criticism from parents for the Ninth Doctor episode "The Unquiet Dead," with its graphic portrayal of zombies roaming Victorian Cardiff in Wales.

But why? What's so scary about a TV show? And if it's really that frightening, then why do we watch it in the first place?

Our feelings of fear are regulated by the brain's limbic system, in particular an almond-shaped region in the temporal cortex known as the amygdala. As soon as the amygdala receives signals warning of a potential danger, it sets processes in motion to help you protect yourself—such as widening breathing passages, increasing the blood flow to your muscles, and stimulating adrenalin production to make your heart beat faster. It also sends nerve impulses to your face that make you grimace and appear fearsome.

These reactions are collectively known as the *fight or flight response.* This comes from millions of years of evolutionary conditioning, in which feeling afraid typically involved saber-toothed tigers and other such nasties.

Clearly, any caveman presented with a saber-toothed tiger would have two options: defend himself (fight) or run away (flight). Either way, if he's to survive then his body must give its absolute best. And that's what the fight-flight response is gearing you up for.

It doesn't take much to trigger a momentary fight-flight reaction. Even a threatening word embedded in a sentence—KILLER—can make the amygdala twitch. And so it's not really surprising that television—with today's large, high-quality screens and surround sound, not to mention the gamut of special effects technologies now available to program-makers—can really give us the willies.

The difference, of course, between *Doctor Who* and an encounter with a saber-toothed tiger is that we know *Doctor Who* isn't real. This is where the brain's frontal lobe comes in. The frontal lobe is involved in a range of functions including judgment, memory, motor control, and problem-solving. But, most important, it's the key to reason and impulse control. The frontal lobe is aware of the bigger picture. It reminds you that although you're feeling fear, what you're seeing isn't real. It tempers the fight-flight response. And it's that feeling of fear while knowing that there's really nothing to worry about that—for some of us, at least—is enjoyable.

Fear and stress also trigger the release of endorphins. These are the body's natural opioids—pain-killers. They're released just in case that saber-toothed tiger actually manages to get its teeth into you. In the absence of a tiger fang in your back, they produce feelings of well-being and euphoria, especially once the cause of the initial fear response is gone.

A more extreme example is bungee jumping. Here, the experience is active rather than passive, and it invades all five of your senses rather than just sight and sound, making it much more intense. You still know deep down that you're safe, but your frontal lobe has to work a lot harder to temper your fear response than it does during your time watching the Doctor. Most bungee jumpers agree, though, that the endorphin buzz they get afterward is well worth it!

18

Exile to Earth

"We're falling through space, you and me. Clinging to the skin of this tiny little world."

—The Ninth Doctor, "Rose"

The Doctor frequently professes his love for planet Earth—his favorite things including, in no particular order, the game of cricket, human beings, and tea. And so it seems rather an odd place for the Time Lords to choose to exile him to at the end of "The War Games"—rather like you or me being dragged kicking and screaming to a beach bar in the Maldives.

As punishment for interfering in the affairs of other races (a serious crime in the Time Lord book), the Second Doctor was sentenced to regenerate his appearance and to be exiled to planet Earth until the Time Lords deemed him ready to leave. In the meantime, the secret of the Tardis was to be taken from him.

But as we saw in Chapter 7, interfering in alien affairs had little to do with it. High inflation in the late 1960s and early 70s meant budgets were tight, and filming a show set on Earth was very much cheaper than trying to recreate elaborate alien worlds.

Young Planet

The planet Earth began forming about 4.6 billion years ago from the gradual agglomeration of dust particles swarming around the young Sun. As the aggregations of particles grew bigger, gravity started to lend a helping hand, and before long boulders and mountain-sized planetoids were smashing into one another, and occasionally sticking together. Soon there were only four large bodies in the inner Solar System—the planets Mercury, Venus, Earth, and Mars.

The surface of the Earth cooled quickly and may have had a water ocean as far back as 4.4 billion years ago. However, the surface was under constant upheaval from volcanic activity, preventing the formation of a stable crust. This hostile environment—made worse by the bombardment from space, as the remaining detritus from the young Solar System was swept up by the planets—continued for a few hundred million years. But by around 4 billion years ago the Earth had become rather more clement, and the first primitive life-forms began to emerge.

There are many ideas as to how this actually took place. The "primordial soup" theory says that the organic materials from which life grew were formed from chemicals present in the oceans and atmosphere of the early Earth. However, this doesn't seem capable of generating the self-replicating DNA molecules needed if life, at least as we know it, is to perpetuate itself. A more promising idea is the "RNA World" theory. This says that prior to the emergence of the first cells, RNA molecules—the molecular cousins of DNA—were the dominant form of life on Earth, carrying out all the biochemical reactions needed for life. This theory shows promise, but scientists aren't yet fully convinced that RNA is capable of replicating itself independently either. Another idea is that life may have been dumped on Earth from outer space—known as the "panspermia" hypothesis. If that's true, then we're all descended from aliens.

In the classic Fourth Doctor adventure "City of Death," life got going rather differently. Here, the primordial soup was jolted into forming the first amino acids for life by the explosion of a nearby alien spacecraft belonging to Scaroth, the last of the Jagaroth. But Graham Cairns-Smith, an expert in the evolution of life at the University of Glasgow, thinks it's highly improbable that life was kick-started by exploding spacecraft or anything else exploding for that matter. "I think the origin of the evolution that led to life on Earth was probably a slow, invisible, and undramatic process—and always very vulnerable," he says. "It's much more likely to be 'kick-stopped' by violent events like this."

Indeed, just such interruptions to the development of life have taken place on several occasions. The geological record of life on Earth is punctuated with a number of mass extinctions, during which the number of species living on the face of the planet dropped abruptly. Geologists refer to the "Big Five" mass extinctions in Earth's prehistory: End Ordovician (445 million years ago, or MYA), Late Devonian (364 MYA), End Permian (252 MYA), End Triassic (200 MYA), and End Cretaceous (65 MYA). The most severe of these was the End Permian extinction, in which 95 percent of marine species and 70 percent of all land vertebrates were wiped out.

Mass extinctions are thought to be caused by a range of different phenomena. These include natural climate change, such as global warming and glaciation. Volcanoes are another cause, which can poison the atmosphere and in extreme cases throw up so much dust that they create a "volcanic winter" blocking out the Sun for months or even years—causing temperatures to plunge and wiping out more species in the process.

In 2005, scientists at the University of Kansas suggested that the End Ordovician extinction may have been caused by a gamma-ray burst. This is a high-energy blast of gamma rays given off in the collision of two neutron stars—the superdense corpses of massive stars. If a gamma-ray burst went off within 30 light-years of the Earth it would irradiate the planet's surface, killing life-forms, and just for good measure would rip away the planet's protective ozone layer as well, making it extremely difficult for life to recover as harmful UV rays from the Sun streamed in.

Perhaps the best-publicized route by which life on Earth can be snuffed out is the impact of a large comet or asteroid. When a body whistling through space at around 30 kilometers per second strikes our planet, the force of the impact is colossal. A 10-kilometer-wide impactor striking the Earth at this speed would release the equivalent energy of around 15 billion Hiroshima bombs. This would trigger fire-storms, giant tidal waves, earthquakes, and an "impact winter" (similar to a volcanic winter) that could last for years. Even small impactors can cause considerable devastation. In 1908, a 50-meter comet or asteroid exploded in the sky over the Podkamennaya Tunguska River in Siberia, detonating with the equivalent force of 10 to 15 megatons of TNT (about 1,000 Hiroshimas) and flattening over 2,000 square kilometers of forest. Had the Tunguska impactor fallen on central London, pretty much everything within the M25 ring road surrounding the city would have been obliterated (see figure 10).

It was almost certainly an impact from space that caused the End Cretaceous extinction 65 million years ago and wiped out the dinosaurs. Geologists are convinced of this because there's a layer of the metal iridium in the fossil record right at the time the dinosaurs died. Iridium is rare on Earth but found in abundance inside comets and asteroids. The geologists even think they've found the site of the crater—a 180-kilometer-wide outline that shows up in geophysical surveys, centered on the Mexican Gulf near the town of Chicxulub on the Yucatan Peninsula. The object that made the crater is estimated to have been around 10 kilometers across and exploded with a force of nearly 200,000 gigatons of TNT (200,000 billion tons).

At least, that's the real-world version of events. In the Fifth Doctor story "Earthshock," things happened rather differently. The Cybermen took control of a giant star freighter, which they intended to crash into Earth to disrupt a conference at which

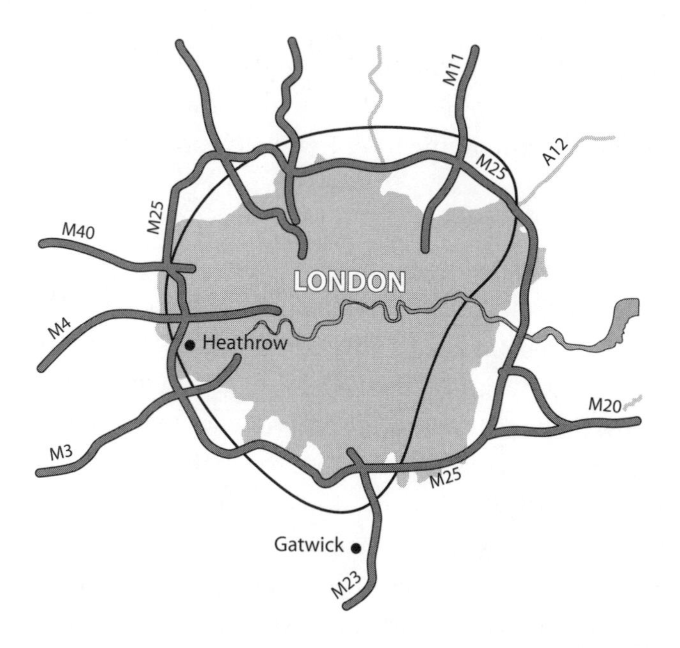

FIGURE 10. The kidney-bean shaped outline shows the footprint of the devastation caused by the 1908 Tunguska impact in which an explosion leveled a large area in Siberia is overlaid onto a map of southeast England. If the impactor had struck central London, almost everything within the M25 road around the city would have been destroyed.

numerous planets were to be united in a war against the Cyber race. The Cybermen had overridden the freighter's controls and protected their override system with three "logic codes," which the Doctor's mathematical genius assistant Adric attempts to crack. He breaks two of them, sending the freighter spiraling back through time in the process. But as he's about to crack the third, a Cyberman warrior regains consciousness and opens fire, missing Adric but destroying the control panel at which he was working. "Now I'll never know if I was right," laments Adric as the ship plunges into the Earth 65 million years ago.

A freighter would have to be awfully large to hit the Earth with as much clout as a 10-kilometer comet or asteroid. But then again, you could argue that a spacecraft like this would have to be packing a fairly heavy-duty propulsion system. If this went up, it stands to reason that there would be trouble to the tune of one seriously big explosion. Presumably, the ship was made of iridium too, to explain the layer in the geological record.

Core Control

Crashing Cybermen starships aren't the only things that have been used in *Doctor Who* to alter the Earth in a big way. In "The Dalek Invasion of Earth," the Daleks dig down into the Earth to try to blow up its magnetic core and replace it with a device that will enable them to pilot the planet through space.

How realistic is it to burrow down to the Earth's core? In May 2003, physicist David Stevenson at the California Institute of Technology figured out how to do just that. In a paper published in the science journal *Nature*, he explained how a small scientific probe could make its way from the surface of the planet down to the core over the course of just a few days.

Forget picks, shovels, and drills. His plan was to open a crack in the Earth's crust—perhaps using a nuclear weapon. Into this, the grapefruit-sized probe would then be dropped, along with 100,000 tons of molten iron (that might sound like an awful lot of iron, but the world's foundries actually turn out this much in about a week). Because the iron is so much heavier than the surrounding rock, it would quickly sink down toward the core, carrying the probe with it and healing the crack behind it as it went.

If Stevenson thinks that he can send a probe to the center of the Earth, then the Daleks can certainly get a bomb down there. Stevenson planned to communicate with his probe using seismic waves, sound vibrations transmitted through the Earth, and perhaps the Daleks could signal their bomb to tell it to detonate in the same way.

A word of caution, though. Swirling currents of molten metal in Earth's outer core are thought to be what generates our planet's magnetic field. As we've seen, the magnetic field fends off harmful radiation from space, so switching it off by destroying

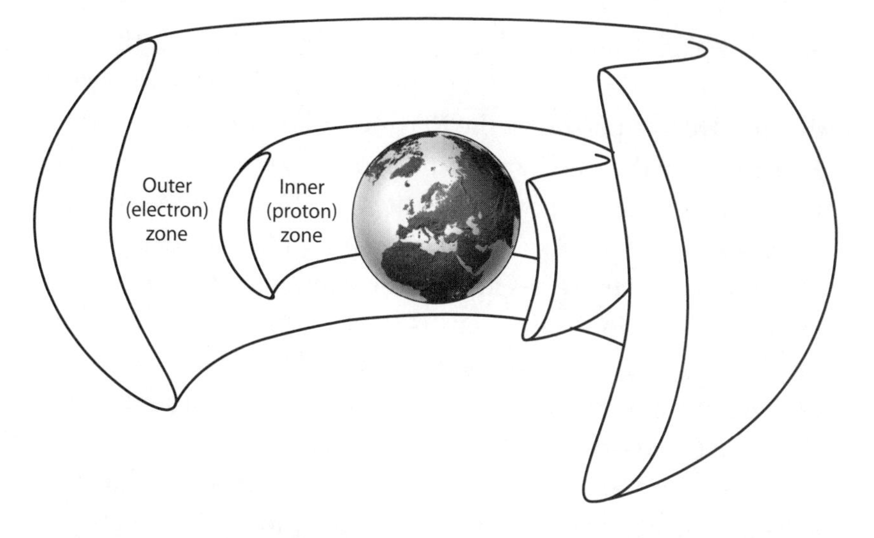

FIGURE 11. The Van Allen belts are two toroidal bands of high-energy subatomic particles circling the Earth, trapped there by the planet's magnetic field.

the core might not be a wise idea. That's especially true in "The Dalek Invasion of Earth" because the Daleks are planning on piloting the planet off into space. The space radiation threat posed by so-called galactic cosmic rays—high-energy particles from space—is strongest outside of the Solar System, where the protection afforded by the sun's magnetic field and the solar wind of particles streaming out from it diminishes.

The Silurians—another of the Doctor's adversaries (see Chapter 13)—tried to attack the Earth from the other direction. They sought to cause calamity not from down below, but from far above: in the Van Allen radiation belts. These are two large, toroidal-shaped belts of high-energy particles from the Sun that have been trapped by the Earth's magnetic field (see figure 11). The belts were predicted by astrophysicist James Van Allen, who was also the first scientist to interpret the data from NASA's *Explorer I* space mission in 1958 as confirming their existence. The belts extend between about 1,000 and 25,000 kilometers above the planet's surface.

In the Third Doctor adventure "The Silurians," the reptilian

Silurians plan to use a "molecular disperser" to dissipate the Van Allen belts and destroy all life on the planet. The trouble here is that the average person on the Earth's surface probably couldn't give a fig whether we have Van Allen belts or not. Unlike the Earth's magnetic field, which creates the belts (and which, as we saw above, does play a key part in keeping the Earth fit for life), the belts themselves are of little consequence. "Removing the radiation trapped in the belts does nothing good or bad to Earthlings, since the radiation is trapped in a place where it does no one any harm," says John Eades, of the European particle physics laboratory CERN in Geneva.

Some critics of human spaceflight have suggested that the belts pose a hazard to astronauts who have to fly through them in the course of their missions. However, the reality is that it would take weeks of living within the belts to accumulate a lethal radiation dose. Nevertheless, the late science fiction writer Robert Forward once proposed a way to drain the Van Allen belts of their radiation by dangling electrically charged cables, or "tethers," into them from orbit. The electrical charge on the cables, he calculated, would deflect the charged particles out of the belts. However, you can't help wondering what the point would really be. The Van Allen belts are replenished by the stream of charged particles from the Sun—so no sooner have you emptied them than they will start to fill up again.

Either way, the Silurians are making dog noises in completely the wrong forest.

Ending It All

Have you ever wondered what will finally bring about the demise of our planet? Global warming? An ice age? The financial crisis? The Ninth Doctor episode "The End of the World" took a refreshingly optimistic view—that Earth enjoys the longest possible life it could hope for, giving up the ghost only as the sun finally expands at the end of its natural life, baking the inner planets to a crisp. The Doctor and his assistant Rose jump forward in time to watch the spectacle from Platform One, a shielded space station orbiting the Earth in the year 5 billion AD.

In fact, 5 billion is a slightly pessimistic estimate for the number of years that the Earth has left. Research by a team of astrophysicists at the University of Sussex has suggested that the heat from the expanding Sun won't render the Earth uninhabitable for about another 5.7 billion years. That's a 700-million-year stay of execution—worth fighting for, I'd say.

The Sussex model is the first to take accurate account of how the Sun loses mass as it gets older—and that's what's bought our planet the extra time. As the Sun racks up the years, it's gradually swelling up, getting bigger and brighter. As it expands, its gravity decreases, making it easier for its fierce radiation to blow its wispy outer layers away into space. The effect is rather like the solar wind blowing away from the Sun today, but much thicker and denser.

Losing mass saps the gravity of the ailing Sun further, making it expand even faster. And as the Sun gets bigger and its gravity dwindles, so its grip on the planets weakens. By 7.6 billion years from now, the Sun will have got as big as it can get—with a radius of 172 million kilometers. However, at the same time the reduction in the Sun's gravity will have increased the orbital radius of the Earth from 149 million kilometers (what it is today) to 185 million kilometers. And so, while the inner worlds Mercury and Venus are engulfed, the Earth itself is spared by a whisker.

That's where the good news ends, however. Although not swallowed up by our expanding star, the Earth will be blasted relentlessly by solar radiation, its surface temperature reaching 2,000°C—hot enough to melt it—by 5.7 billion AD.

The next planet out from Earth—the cold, currently uninhabitable world Mars—will become quite temperate 7 billion years from now as the Sun expands farther. However, this period of clemency will last for only 100 million years or so (not even long enough for primitive bacteria to evolve) before the temperature soars and the red planet is incinerated as well.

Roughly 7.75 billion years from now, the Sun itself will bow out, flinging off its outer layers to leave a white dwarf star. The white dwarf is the remnant state of the Sun's core, and is about the same size as the Earth. Although it has no power source of its own, it's fiercely hot—100,000°C—and its small size means that it cools slowly, so that it will glow brightly for tens of billions of years to come.

Some astronomers say they've found evidence for planet-like objects in orbit around some white dwarfs. And this has led scientists to speculate that some of our Solar System's planets—perhaps even the charred remains of the Earth—may live on into eternity.

19

The Human Empire

"The Fourth Great and Bountiful Human Empire. Planet Earth is at its height, covered with megacities, five moons, population 96 billion, the centre of a galactic domain that stretches across a million planets and species."

—The Ninth Doctor, "The Long Game"

Humans—and all other terrestrial life-forms, for that matter—aren't going to be living on planet Earth for ever. As we saw in the last chapter, even if our world survives the numerous natural calamities that it faces, like asteroid impacts and nearby stars going supernova, then the Sun has a bigger surprise in store for us when it eventually expands and roasts the Earth to a crisp.

So at some point humans are going to have to think about leaving the Earth behind and spreading out across one or more other worlds. Space empires have often featured in *Doctor Who*—see, for example, "Mission to the Unknown," "The Daleks' Master Plan," and "The Dominators." In "The Curse of Peladon," we learn about the Galactic Federation—an interplanetary organization spanning much of the galaxy. And the Ninth Doctor episodes "The Long Game," "Bad Wolf," and "The Parting of the Ways" introduce us to the "Fourth Great and Bountiful

Human Empire"—the future of human civilization, spanning hundreds of thousands of planets.

Assuming we can launch ourselves into space without too much difficulty—which seems a reasonable assumption for the far future—is it possible that humans might one day build empires such as these, on interplanetary and perhaps even interstellar scales?

Striking Back

Seth Shostak, a senior astronomer at the Search for Extraterrestrial Intelligence Institute, is doubtful. He draws a parallel with historical empires on the Earth. Take the Roman Empire, which extended over most of the Mediterranean and up through Europe to Hadrian's Wall in northern England. "They were able to do that because communication times were on the order of a few weeks," says Shostak. "So, say if the Germanic hordes had decided to cross the Rhine again at Mosel, a runner would be sent down to Rome. The guys there would make a decision, they'd move a couple of legions into place, and they could usually throw the rebels back onto the other side." That typically happened on timescales of weeks, or maybe a month or two at the most.

It was a similar case with the British Empire. This was able to extend to the whole globe as better technology enabled faster communication and transport. But the timescale on which the Empire was able to respond to threats was still on the order of weeks or months.

This is the snag, because when you try to spread a civilization out across a galaxy, suddenly the timescales involved become very much longer. Let's take the Fourth Great and Bountiful Human Empire of the Doctor Who universe. We're told that this spans a million planets. There are hundreds of billions of stars in our galaxy; so, assuming there's on average one habitable planet per star, then the Human Empire takes up about one hundred-thousandth of the volume of the galaxy. Spiral galaxies like our own typically measure roughly 100,000 light-years across by about 3,000 light-years thick. So that means that humans have colonized a spherical region in the galactic disk about 800 light-years in diameter.

Now imagine that the Daleks have decided to attack a small colony in, say, sector 9 of the great Empire. Outnumbered, the colonists send an SOS message back to Earth, but because of the light-travel time this takes hundreds of years to get there. As soon as the Federation receives the message they send a battle-cruiser but, according to current physics, that can't go any faster than lightspeed, so at best it'll take another 100 years to reach sector 9. "By this time, whatever the Daleks had in mind to do—they've already done it," says Shostak. "In fact, everybody in sector 9 is a Dalek!"

In fact, the situation is even worse. Our initial, rough assumption of one habitable planet per star is, in fact, very optimistic. The true figure is probably a lot lower, making the Fourth Great and Bountiful Human Empire much bigger (in order to encompass the million planets we're told it contains), and so the communication timescales across it are even longer still.

As we'll see in Chapter 25, new physics may make it possible to break the speed of light and so alleviate the travel-time problem. But high-speed travel opens the door on a whole raft of new problems for a space empire. When an astronaut travels off at high speed, the clocks in his or her frame of reference slow down. We encountered this "time dilation" phenomenon when we talked about time travel in Chapter 3. Here's the problem. Imagine the Ninth Doctor sends Captain Jack Harkness off on a mission from Earth to the Cyberman planet of Telos. From Captain Jack's perspective the round trip takes, let's say, just a year. But if his spacecraft travels at 99.999 percent of the speed of light to do this, then time dilation means that his clocks are running roughly 200 times slower than the Doctor's. So when Jack returns, the Ninth Doctor is 200 years older—by then, he's probably the Twelfth Doctor.

Now imagine that the Federation has a whole fleet of starships all doing this. Are the Federation commanders really going to be able to coordinate all this with every clock on every starship showing a different time and running at a different rate? "The synchronicity of things gets messed up very easily if you have travel close to the speed of light," according to Shostak.

So it seems, realistically speaking, that we could sustain an empire in space that's only light-weeks, or at the most a few light-months in size. "You might just be able to hold something together over the distance to the nearest stars, which are a few light-years away," Shostak states. Asking to be the Emperor of anything bigger, however, could be asking for trouble.

New Civilizations?

That would seem to limit the number of planets in any interstellar empire to just a handful—not much of an empire to speak of at all. And in terms of creating more real estate for when we run out of space on Earth, it's a disaster. The surface area of any new planet is going to be comparable to the area of the Earth. So colonizing that planet will only double the amount of land we have available to build on. That sounds like a lot of land, but Earth's population itself is doubling about once every 40 years (the global population was around 3 billion in 1960; it's now around 6.7 billion). So moving to Mars when the Earth fills up is going to buy you only a few more decades—all the effort doesn't really get you very much.

Perhaps a more sensible option is for us to live in giant space stations (see Chapter 26), or even to colonize the asteroid belt between Mars and Jupiter. This latter possibility was suggested by the eminent Princeton physicist Freeman Dyson. There are a lot of asteroids, and so plenty of acreage to build on. And asteroids have all the ingredients—metals, minerals, and so on—that we'd need to build cities.

Of course, those are the purely technological issues. That's to say nothing of the sociological ones. Take a look at any empire on Earth, and it's the human factors that usually decide its future, or rather its fate. "Look at what happened when England spread to North America," says Shostak. "Everything was fine and dandy from 1607 to 1776—but that's less than two centuries!" Even the mighty Roman Empire only managed a little over 500 years. It seems likely, then, that after a few centuries a colony on another planet will become a different society and won't necessarily want to be part of the empire any more. Then it'll be time for Martian independence day.

That's not to say that in the centuries and millennia to come, human colonies won't spread across the galaxy. In fact, it will be necessary if our species is to survive. We've had our eggs in one basket for far too long already, and migration to other worlds will be our insurance policy against the Earth meeting an untimely end with all of humanity still on it.

But will these new worlds and new colonies form some kind of grand human empire? Probably not; definitely not for very long. As the Seventh Doctor put it in "The Happiness Patrol": "I can hear the sound of empires toppling." Because that's exactly what's happened to every one that's been built so far.

20

Invasion Earth

"Jenkins!? Chap with the wings there—five rounds, rapid!"

—Brigadier Lethbridge-Stewart, "The Daemons"

Even if empires are unlikely, as we saw in the last chapter, that doesn't rule out conquest in space. Not all wars are fought in order to build up a cosmic commonwealth. So if there are intelligent aliens living on worlds dotted across our Milky Way galaxy, will they all want to be our friends or will some of them be hostile? And what will we do when the rough crowd from Alpha Centauri finally turns up on our doorstep?

The idea has frequently been aired in *Doctor Who.* Indeed, alien invasion is the show's bread and butter. Most Dalek or Cyberman adventures involve the villains using force to be somewhere they shouldn't—often planet Earth. And plenty of other alien races have taken a crack at us over the years, including Sontarans, Ice Warriors, robot Yetis, the Autons, the Slitheen, and more recently the Sycorax.

"I suspect that in a Universe full of life, some extraterrestrials are likely to be friendly, some hostile, and some neutral," says Nick Pope, who once investigated the UFO phenomenon for the British government's Ministry of Defence.

Some researchers are so convinced that there are hostile aliens out there, they think we should be careful about announcing our presence to the rest of the galaxy. Whether that's true or not, it's a bit late to be complaining now. We've been beaming radio signals out into space—in the form of our TV and radio broadcasts—for more than half a century. So anything within 50 light-years of the Sun already knows we're here, and is right now probably trying to make sense of *American Idol*—and plotting its revenge.

In 1974 Frank Drake, now Emeritus Professor of Astronomy and Astrophysics at the University of California, Santa Cruz, beamed into space one of the first direct signals to an alien world. He used the 300-meter Arecibo radio telescope to send a message—containing basics of number theory, chemistry, and biology and a picture of the Arecibo dish itself—to M13, a globular star cluster in the constellation of Hercules. M13 is 25,000 light-years away, so don't expect a reply any time soon. But Drake argued that there was no danger in sending his message. He believed that if a civilization is to survive long enough to become technologically advanced then it must also overcome its warlike tendencies and become altruistic rather than malevolent. Otherwise, he says, it will destroy itself.

Of course, even if there are hostile aliens in the galaxy, it's only a problem if they have the means to come over here and attack us. Could other civilizations have mastered the art of interstellar travel yet? Pope thinks they have, and that we're already playing host to extraterrestrial visitors—though this view is by no means universal. In fact, many scientists look on the field of UFO research with extreme skepticism. Pope admits that most UFO sightings can be explained in terms of misidentifications of known objects or phenomena or as hoaxes. He concentrates his efforts on the small percentage of reports that he believes cannot be written off in this way. "There have been thousands of sightings by trained observers such as police officers, pilots, and military personnel," he says. "On numerous occasions, visual sightings have been confirmed by radar."

But Seth Shostak of the Search for Extraterrestrial Intelligence isn't convinced. "It's safe to say these witnesses have seen something," he says. "But just because you don't recognize an aerial phenomenon, it doesn't mean that it's an extraterrestrial visitor. That requires additional evidence."

As we saw earlier, the vast distances between the stars will certainly make the journey of any invading alien fleet a long and arduous one. The *Voyager 2* space-probe, launched from Earth in August 1977, is currently on its way out of our Solar System, but it's not going to reach the nearest stars for another 300,000 years. Those sorts of timescales are typical. Even Pope, who thinks that aliens are already able to make the journey, admits that no beings of a hostile nature have managed to get here yet.

But when or if they do arrive, what are we going to do about it? After all, a hostile alien race that has crossed space to reach us must either be highly advanced or highly determined, or worst of all: both.

"I used to run the British government's UFO Project, but we weren't set up along the lines of UNIT," says Pope. Run by the Doctor's colleague Brigadier Lethbridge-Stewart, UNIT—the United Nations Intelligence Taskforce—was an organization set up to investigate and combat extraterrestrial (and sometimes paranormal) threats to the Earth. (Readers of a more geeky persuasion might like to visit the UNIT website, www.unit.org.uk. Browse the site for the basic experience, or log in using the Doctor's personal password, "buffalo"—as revealed in the Ninth Doctor episode "World War Three"—for enhanced content.)

Pope isn't aware of any organizations like UNIT anywhere in the world. So if a Cyberman invasion force turned up tomorrow, it would be up to the regular army to defend us. Given the space transportation problem the aliens would face, we might not be outnumbered, though their superior technology would probably leave us outgunned.

Maybe we'll shoot back with all that alien technology we've been stockpiling at Area 51 and other secret warehouses around the world (like Van Statten's underground bunker in the Ninth Doctor episode "Dalek")? Or then again, maybe we won't. Area 51 certainly exists—it's part of the Nellis Air Force Base in Nevada. And, yes, it's used for testing future aircraft designs—the sort of stuff that won't be seen at the Paris or Farnborough air shows for another 10 to 15 years. "But none of this material is alien in origin," says Pope. "If it was, NASA would be doing interstellar missions by now. Instead they're struggling to even fly space shuttles."

Game Over

Does this mean we're done for? Not necessarily. You wouldn't expect an alien fleet to trek halfway across the galaxy just for target practice. The chances are that if or when they arrive they'll be here because they want something from us, and so we can realistically hope that they're not going to simply reduce the Earth to smoldering slag from orbit. In that case, it seems likely that we'd be fighting at a tactical rather than a strategic level. And that could mean that we're in with a fighting chance. Humans aren't exactly unskilled or unpracticed in the art of war and, of course, we'll have the home field advantage on our side. One only has to look at the heavy casualties sustained by the American military in Vietnam to get an idea of how dangerous it can be to corner even a technologically inferior enemy on his own turf—especially when that turf is so markedly different from your own.

If "fighting us on the beaches" isn't going to guarantee them success, might the aliens try slightly more insidious tactics? They already are, if some witnesses are to be believed. Many people who have reported UFO sightings also claim to have had alien abduction experiences, in which they have been taken against their will aboard an alien spacecraft, and in some accounts even been experimented on. In 2003, it was estimated that some 4 million Americans believed they had been the victim of an alien abduction.

Psychological studies of "abductees" show that many of them exhibit real stress symptoms when they tell of their experiences, similar to those exhibited by battlefield veterans. "The results suggest that the abductees aren't lying—they genuinely believe they've had these experiences," says Pope.

However, Richard McNally of Harvard University, the psychologist who conducted this study, argues that his findings are just a symptom of misguided or delusional emotional belief. "If you genuinely believe you've been traumatized and recall those memories, you'll show the same psycho-physiologic emotional reactions as people who really have been traumatized," he says.

Some parapsychologists believe that alien abduction accounts could be the product of what's called "sleep paralysis." This condition is caused as the brain enters a state of wakefulness while the body is still experiencing the paralysis that affects it during sleep. The result can be auditory and visual hallucinations and even feelings of dread and sensing a "presence."

You could accuse the Doctor of alien abduction himself—for all those human assistants that he's made off with over the years. He was nearly turned over to the police by Rose's mum for that very reason in the 2005 series. Then again, most of his companions have signed up willingly. They know what they're letting themselves in for. And, after all, it won't be the end of the world. Will it?

• Part Three •

ROBOT DOGS, PSYCHIC PAPER, AND OTHER CELESTIAL TOYS

21

Scanning for Alien Tech

"I think you should do a scan for alien tech. Gimme some Spock! Just once, would it kill ya?"

—Rose, "The Empty Child"

What's the best way to see an elephant? Go to the zoo and look for one, or sit by the phone waiting for it to call? Now researchers looking for the signs of alien life in the Universe are cottoning on. Rather than listening for deliberate attempts by aliens to communicate with us, they're planning to scour the galaxy for the exhaust trails of alien spacecraft and the shadows made as alien space stations glide in front of their home stars.

They call their project the Search for Extraterrestrial Technology (SETT). According to the Autumn 2005 issue of *BBC Focus* magazine, Luc Arnold at the Observatory of Haute-Provence, France, thinks that it will be possible to use the data from a NASA space telescope to scan the skies for the signatures of alien engineering projects. The Kepler space telescope, which launched in March 2009, is tasked with monitoring 100,000 stars for the dips in brightness caused as planets orbiting around them cross in front of the stars' bright disks. Arnold believes that the telescope can do more than simply detect changes in brightness, though. He says that the probe will be able to tell the difference between the round shape of a planet and the kind of angular, slotted outline of an artificially constructed object.

True, the structures would need to be fairly big—about the size of a planet. But astronomers working on SETT have speculated that the objects detectable by the Kepler may be just the tip of a fairly huge iceberg. Other detectors, such as the new generation of ground-based telescopes with mirrors a staggering 100 meters across (or bigger than an American football field), could be able to spot alien solar sails (spacecraft powered by giant reflective sheets that hitch a ride on starlight—see Chapter 25), alien nuclear waste dumps, and the exhaust from any antimatter-powered starships that might be out there.

Some even think we might find evidence for alien hardware closer to home. They've pointed out how the Moon acts as a gravitational sponge, sweeping up objects from space that pass near to it. Of course, any massive body will do this. But objects landing on Earth are soon eroded away by the planet's relentless atmospheric and weather systems. What's special about the Moon is its lack of an atmosphere, making lunar conditions much kinder, so that objects deposited there do not erode. That's not to say that when NASA returns to the Moon in 2019 they're going to find Martian iPods lying around. But it's been suggested that we should search the Moon for materials that are of an obviously artificial origin.

So what have the *Doctor Who* writers made of alien technology? Of all the eleven Doctors so far, the biggest fan of gadgets and technical tricks by a long way was Jon Pertwee's Third Doctor, especially during the time of his exile to Earth. When he wasn't busy reversing the polarity of the neutron flow (and, yes, only a true pedant would point out that neutrons are called neutrons because they're electrically neutral and thus have no polarity to reverse, so we won't get into that), he was occasionally seen buzzing around in the Whomobile—a flying car built by UNIT that appeared in "Invasion of the Dinosaurs" and "Planet of the Spiders."

Technology pundits have been telling us that flying cars are but a few years away for decades. But now, they may have got it right. California-based firm Moller International has developed what it calls the world's first "personal vertical takeoff and landing vehicle." Called the M400 Skycar, it's a four-seater car-sized aircraft powered by eight high-power ducted fan engines—a propeller mounted inside a moveable duct—arranged in four pods around the craft. The top speed of the vehicle is 375 mph, with a cruising speed of 275 mph. Fuel consumption of the M400 is around 20 mpg and the range is about 1,200 kilometers (750 miles). Moller says that a prototype of the Skycar will be in the air by 2012, for an initial cost of around $500,000 each. The company hopes to bring the price down to $60,000 eventually.

Even if you have that kind of money kicking around, you'll also need a pilot's license. However, the company states that it intends for the Skycar to ultimately be able to fly on full autopilot, so that no skill will be required on the part of the operator. The good news is that the M400 does come with parachutes capable of bringing the whole thing safely back down to the ground in the event of an emergency. The bad news is that, at the moment, the law says you can land the things only at designated airports and heliports, so using them to do the shopping is probably still some way off.

When he's not fiddling with his own gadgets, the Third Doctor encounters some bizarre pieces of alien kit in the course of his travels. For instance, in "Carnival of Monsters," he ends up inside a miniscope—a kind of portable, miniaturized zoo containing environments from various planets, populated by real, miniaturized creatures. In the adventure "Nightmare of Eden," the Fourth Doctor has to deal with a similar piece of technology known as a "continuous event transmuter," which stores chunks of planets on recordable crystals.

Could devices like this work? Sadly, the answer is "no." There's a detailed discussion as to why miniaturization is currently thought to be impossible in Chapter 11.

However, it could be possible to make recordings of an environment that are detailed enough to construct a convincing 3D copy from. Robert Freitas of the Institute for Molecular Manufacturing in Palo Alto, California, points out that the entire surface area of the Earth is about 500,000 billion square meters. If you map this to the resolution of the human eye, which is about 30 microns (30 thousandths of a millimeter), and to a depth of about 100 meters so you retain all the visible 3D information, and if you then assume that each pixel in this map will use up about 6,400 bits of memory (100 bytes for a 64-bit processor), then it'll take about 10 million billion billion billion bits of memory to map the entire planet.

How small a volume of space could all that information be stored in? Ignoring technological considerations for a moment, the maximum density of information allowed by physics was calculated in 1974 by Jacob Bekenstein, currently Polak Professor of Theoretical Physics at the Hebrew University of Jerusalem. According to Bekenstein's figures, all of the information needed to make a virtual Earth could be packed down into a volume of just a cubic centimeter and so could easily be fitted into a handheld device.

Back in the real world, where we're constrained by technology, this number's going to be rather bigger. Freitas says that the highest-density information storage will be attainable using molecular nanotechnology—building machines out of molecules on scales of a billionth of a meter.

The concept was first put forward by American engineer Eric Drexler in the early 1980s. Now engineers are becoming increasingly proficient at manufacturing components on such minuscule length scales, including cogs, paddle-wheels, and even artificial molecules. As long ago as 1995, a British team of chemists built a molecule consisting of five interlocking rings in the shape of the Olympic logo—each ring was made up of 75 atoms. Ralph Merkle of the Georgia Tech College of Computing once said that, next to molecular nanotechnology, conventional manufacturing is like trying to make things out of Lego bricks while wearing boxing gloves on your hands. Over the coming years, researchers expect to be able to assemble tiny components into intricate nanoscale machines, which Freitas refers to as nanorobots.

Freitas says that molecular nanotech will enable storage capacities of around 10 billion bits per cubic micron. So to store enough data to make a "copy" of the Earth will require a box that's a million cubic meters in volume or 100 meters along each side.

If you wanted to record your world right down to atomic resolution—so that each pixel is about a nanometer across (a billionth of a meter), then the storage requirement goes up by a factor of 30,000 billion. "This would mean a Bekenstein-bound data bank volume of 27 million cubic meters—a cube 300 meters on an edge," says Freitas. "Which would probably be pushing your luck, even inside the Tardis!"

If you had such a recording, how might you go about watching it? It would be a simple matter for a computer to take the data and display any kind of view you might desire on your screen—a kind of ultimate Google Earth. But it would be much more dramatic if the data could be rendered in some kind of 3D environment that the user could actually walk through—a kind of holographic image projector. The Fourth Doctor met such things in "The Pirate Planet" and also in "The Leisure Hive," in the form of the "recreation generator"—a device that was able to project solid images for recreational activities.

Freitas thinks that such devices might also be possible through nanotechnology. He imagines a swarm of nanoscale robots, able to configure themselves into complex surfaces of any shape, size or texture. He calculates that just a kilogram of these nanorobots would be enough to construct a 10-micron-thick programmable sheet, 50 meters square—basically a "smart screen" that can be morphed into any shape you like and then have images displayed on it. He admits that nanosurfaces representing complex forms like people and animals would need to be more elaborate. They would probably require extra "ballast mass" to make them convincing up close and to shape the internal organs needed to enable more involved activities such as speech and eating.

Much the same technology can be invoked to explain a device called the "transmogrifier," from the Sixth Doctor adventure "Vengeance on Varos." This device alters a person's appearance until they take on the form that they subconsciously aspire to. The Doctor's assistant Peri, for example, is temporarily transformed into a birdlike humanoid. We're told that the transmogrifier was developed for mining research, so that miners could grow claws to help them dig.

Freitas thinks that this could work with the help of a suit of nanorobots that the user would wear. "If these nanorobots were under the person's control, he or she could program them to assume whatever external shape they desired," he says. "In effect they would form a whole-body 'mask' around the unchanged human body."

There would be limitations on the size range that the transmogrified shell could take. Most importantly, it would have to be at least as large as the person inside—simply so that they didn't get squeezed to death. But also, it shouldn't be too large, otherwise serious problems with balancing the extra load and even just carrying it all around could soon result. Strands of nanorobots could easily be joined up to simulate hair, fur, and even feathers—so Peri's transmogrification into a bird would be quite possible within this scheme.

As an aside, some amount of transmogrification may be possible without the help of any technology at all. "There is some very obscure literature on people taking on the appearance of who they think they are over time," says Richard Wiseman, a psychologist at the University of Hertfordshire in England. In particular, one research paper shows that children who are more deceptive at a very young age start to resemble our typical images of how dishonest people look.

"Jiggery Pokery"

The simplest, most trivial pieces of technology are often the best. In the Ninth Doctor adventure "The End of the World," the Doctor performs what he refers to as a little "jiggery pokery" on his homesick assistant Rose's cell phone to enable her to call her mother from the year 5 billion AD.

Clearly, this isn't a service she's going to get with Verizon any time soon. There's no way old-fashioned electromagnetic waves can send a signal into the past like that. But theoretical physicists might just have something else that can. So-called tachyons (see figure 12) are subatomic particles that travel faster than the speed of light. Einstein showed that if a particle can travel faster than light in one frame of reference, then you can always find another where it travels backward through time. Then, just as broadcast engineers can piggyback a signal on a beam of radio photons to transmit the signal through space, so tachyons could be used to piggyback a signal into the past.

The speed of light is often thought of in Einstein's theory of special relativity as a maximum upper speed limit. But it's better to think of it as a boundary that cannot be crossed. So whereas normal matter can never accelerate to the speed of light or above, tachyons can never slow to the speed of light or below. In fact, when a tachyon loses energy it accelerates. The lowest energy state of a tachyon has the particle moving through space at infinite speed.

But could these particles be real? Tachyons are predicted by some speculative theories of particle physics, such as exotic versions of string theory. However, there is as yet no solid physical evidence for their existence, and—alas for Rose—most physicists are doubtful that there ever will be.

So phoning the past looks set to remain troublesome. However, phoning other countries—at least, those where English isn't the native language—could be about to get a lot easier. On several occasions the Doctor has encountered alien races using electronic language translators to talk to one another. For exam-

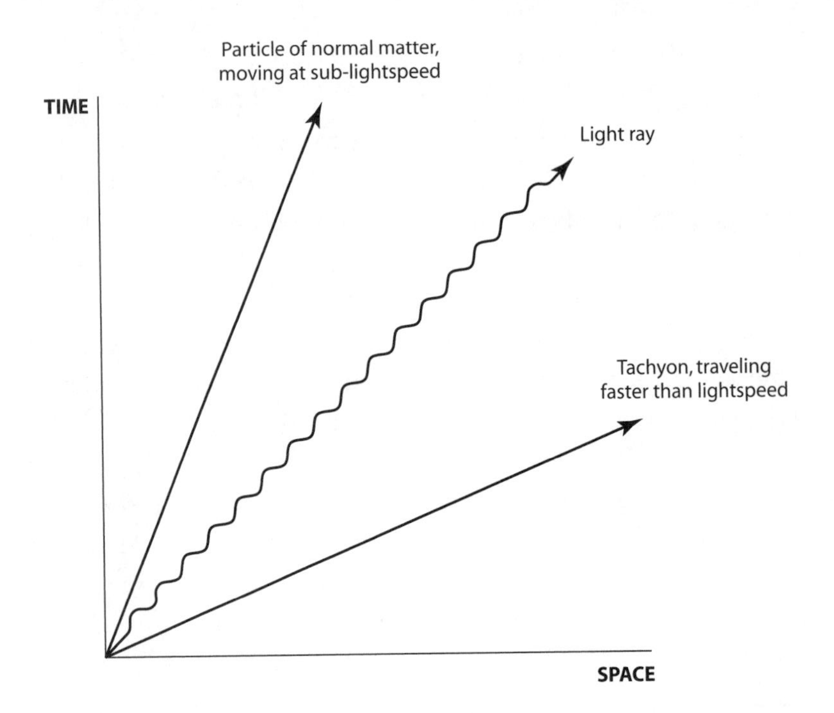

FIGURE 12. Tachyons are particles that travel faster than the speed of light, as shown in this diagram. Einstein showed that such particles can be made to travel backward through time.

ple, the sluglike alien Sil in Vengeance on Varos communicates using what he calls a "language transposer." And the Sontarans, when encountered by the Third Doctor in "The Time Warrior," carry language-translation devices on their belts—presumably so they can bellow comprehensibly at their enemies.

These devices pick up what an alien says, translate it, and then output the result to a speech synthesizer, all in real time. Anyone who has used computer speech-to-text software will either be rolling their eyes or laughing hysterically at this point. It's true—existing systems aren't that impressive. But most computer scientists seem agreed that this is something that is due to improve.

"It's entirely possible," says Steve Grand, an independent computer scientist based in Flagstaff, Arizona. "We just don't quite know how to do it properly yet." Grand thinks the big problem will be getting translation machines to really understand what it is they're translating. After all, how else will they deal with quirks of language such as local idiom? For instance, Grand picks on the French *une nuit blanche.* Translated literally, this means "a white night." But when the French use this phrase, what they actually mean is "a sleepless night." Once machines can acquire some degree of artificial intelligence, which computer scientists are currently working on, they will be able to translate phrases such as this correctly.

Of course, you don't always have to open your mouth to communicate—you can say an awful lot about yourself just through the clothes you choose to wear. In the Second Doctor adventure "The Macra Terror," we meet a handy piece of equipment to help you keep your togs natty. Called the "clothes reviver machine," you simply get into it for a few moments and emerge with your clothes cleaned, pressed, and looking as good as new. How useful would one of those be in the morning?

In 2002, Surrey-based inventor Jonathan Nwabueze came forward with a design for something similar. Called the boardless iron, it was touted to take the drudgery out of ironing. It works using a fan powered by an electric motor to draw the cloth toward a heated plate. Steam is generated in a chamber above the plate, which is then blown onto and through the fabric by the fan to remove creases. Garments are generally ironed while they're hanging up, although "it can even smooth out wrinkled clothes while they are being worn," said Nwabueze. Alas, that was in 2002 and nothing resembling this has, as yet, appeared in the hallowed pages of the Ronco catalogue.

Dress to Egress

Dressing to impress is one thing. But when your destination is the chilly vacuum of outer space, your garments need to be made of sterner stuff. NASA's Extravehicular Mobility Unit—that's a spacesuit to you and me—weighs around 135 kilograms and serves as an astronaut's uniform, workplace, and toilet in the darkness of space, for anything up to 7 hours at a time. It's made up from multiple layers of material. The inner layer, called the bladder, is molded from urethane-coated nylon and keeps the suit pressurized. Over the bladder is a strong layer of Dacron called the restraint, which stops the stretchy bladder from expanding like a balloon and bursting. On top of this is a layer of Ripstop nylon to prevent the suit tearing, no fewer than five layers of reinforced Mylar, and finally an outer fire-resistant coating made from, among other things, Gore-tex and Kevlar. These tough outer layers protect the astronaut from micrometeorites, which, although only dust-sized, can pack as much energy as a speeding bullet. Each NASA spacesuit is made up of almost 20,000 components and costs over $10 million.

So you have to wonder whether the Doctor is being entirely wise when he ventures out into space with just a space helmet and an air supply in the Fifth Doctor story "Four to Doomsday." Although representations of people exploding and their eyes popping out aren't true to life, it's fair to say that a couple of minutes in space without a suit would be enough to kill most humans (and probably Time Lords too). In 1965, a spacesuit test at NASA's Johnson Space Center went wrong, and the test subject was accidentally exposed to a hard vacuum. He lost consciousness after 14 seconds, later saying that the last thing he remembered was the sensation of the saliva boiling on his tongue (lowering the atmospheric pressure lowers the boiling point of liquids). Having oxygen to breathe is all well and good, but the fact is that when you're exposed to a near-vacuum all the gas is going to be sucked out of you anyway—and any attempt to hold air in your lungs isn't advisable as it will cause them to burst.

Breathing can also be difficult in some environments on Earth, such as up a mountain or under water. But it's fair to say these environments aren't quite as harsh as outer space and so here there's no need for a full suit—a tank of air will do nicely. In the First Doctor serial "The Web Planet" we're introduced to the space-age alternative: breathing pills, which the Doctor gives out to his companions to help them breathe in the planet's thin atmosphere.

"These could be made from erythropoietin," says biologist Peter Barlow. The glycoprotein hormone erythropoietin, or EPO, is a growth factor that boosts the production of oxygen-carrying red blood cells. Indeed, it's already used as a doping drug in endurance sports because of its ability to boost oxygen transport around the body.

Robert Freitas has taken this a step further, designing what he calls the respirocyte—an artificial red blood cell. Constructed using techniques from nanotechnology, the respirocyte comprises a 1,000-atmosphere miniature pressure vessel, capable of delivering 236 times more oxygen than its biological counterpart.

"My writings have described respirocytes as injectables," says Freitas. "But you could equally well put them in swallowable pill form and give them some mobility to enable them to get into the bloodstream quickly."

Innovations such as these won't just be used to further the careers of dishonest athletes. There will be obvious benefits to firefighters and military personnel and to anyone working at high altitude or where oxygen levels are low—exactly the circumstances in which breathing pills featured in *Doctor Who*. There will also be clear medical applications, assisting patients with breathing difficulties. In fact, research such as Freitas's, as well as pioneering new developments in surgery, promises a bright future indeed for medicine, with some amazing new developments in store. Many of these have been previewed already for us by the *Doctor Who* writers. And that's why they're the topic of the next chapter.

22

Just What the Doctor Ordered

"An apple a day keeps the . . . Ah, never mind."

—The Fifth Doctor, "Kinda"

As applications of new technology go, medicine rates among the most crucial. Saving lives and curing illness are two of the fundamental motivations driving scientists to make and apply new discoveries. That means that any science fiction hoping to paint an accurate picture of advanced technology in the years, centuries, and millennia from now needs to speculate on where medical technology is likely to go.

Doctor Who has been no exception. From the exploits of the First Doctor, and his use of the "fast-healing bandage" in "The Edge of Destruction," through to the Ninth Doctor and the Chula nanogenes (see below) in "The Empty Child" and "The Doctor Dances," the development of medicine in the future and on other worlds is something that the Who writers have gone to considerable lengths to address.

The Little Things

The nanogenes were an army of tiny robots, each between a few hundred and a few thousand nanometers (billionths of a meter, abbreviated nm) in size, which can work on the body on the smallest scales to repair tissue damage and fight off infections. In "The Empty Child," the nanogenes heal a rope burn on the hand of the Doctor's assistant Rose in seconds.

Robert Freitas of the Institute for Molecular Manufacturing has already designed something very similar to the nanogenes of the *Who* universe. "A few years ago, I designed some nanorobots that would be suspended in a fluid that you'd put on your cut finger, and the cut would be fully repaired in about 1 minute," he says. That's for a shallow, clean cut. Deeper wounds, or wounds with ragged edges, might take a little longer—perhaps on the order of a few hours. The patient probably wouldn't feel any discomfort as the nanorobots went to work—any more than we feel any discomfort as bacteria do their stuff inside us (and bacteria are roughly the same size as nanorobots).

Once they'd made their repairs, the nanorobots would simply collect in a little pile on the surface of the skin where the cut used to be, along with whatever detritus they'd debrided from the wound during repair operations. They could then be collected and reused.

As well as being applied directly to cuts and other external injuries, they could also be ingested or injected to rectify internal problems. Once they'd done their job, the nanorobots would be programmed to swim to the patient's kidneys, from where they would be excreted in the urine. An advanced toilet facility might then be able to filter the urine and extract the nanorobots so that they could be reconditioned and used again.

In some scenarios, where nanorobots are meant not just to heal but also to actually upgrade the functionality of the human body, the robots might stay in the body permanently. They would give your healing mechanism a constant boost, perhaps keep your aging process in check, and also monitor for the early signs of the onset of serious illness, such as cancer—and maybe even take steps to combat it. In this case, you would probably need to top up your nanorobots with a fresh supply of new machines every now and again as the old ones wore out. However, there are also problems of "biocompatibility"—making the robots work seamlessly alongside human biochemistry—that must be addressed before this kind of long-term nano-augmentation of the body can be applied in practice.

Freitas says that his robots, which would probably be made from a durable form of carbon, such as diamond or graphite, could be within the realms of real technology by the 2030 time-frame.

In "The Empty Child," the nanogenes had arrived on Earth in London during the Second World War, as part of the medical facilities aboard a warship belonging to the Chula race. The nanogenes had never before encountered a human being until they came across a little boy wearing a gas mask. From that moment on, they assumed that the gas mask was part of the human form, and whenever they "healed" anyone the patient would subsequently grow a gas mask from his or her face. Realistic?

Freitas thinks that this scenario is unlikely. "A device that cannot distinguish between living and non-living matter, or between the patient and the patient's clothes, is a poorly designed and programmed system that I would judge to be entirely unsafe to be allowed to perform any kind of important medical procedure," he says. "Aliens aren't that stupid."

Chop and Change

The neat, painless techniques of nanomedicine seem a far cry from the messy business of surgery. During the Fourth Doctor story "The Brain of Morbius" we get a glimpse of where surgery may go in the years to come. In a twist on the old Frankenstein tale, the Doctor meets the mad scientist Mehendri Solon, who is assembling a new body to house the brain of the executed Time Lord criminal Morbius.

According to John Fabre of King's College London School of Medicine, surgically speaking you can transplant almost anything. Transplanting a brain is certainly possible—in as far as the recipient could survive the operation. The trouble lies in making the transplanted brain function in its new host. Performing a brain transplant would naturally mean severing the spinal cord, and at present there's no way to repair such a spinal injury. This means that any recipients of such a procedure would be left quadriplegic for the rest of their life, unable to move any of their limbs. And they would be dependent on life-support machines, as the upper part of the spinal cord is needed to operate the lungs in breathing.

"There are also issues of linking into the nerves of the head like the eye, ear, hearing, and so on," says Fabre. And, despite a huge amount of research into the area of regrowing central nervous system tissue, such as the spinal cord and the optic nerve, there seems to have been little real progress. "It's impossible to say whether it'll ever be solved."

A brain transplant has never actually been performed, but in 1970 Robert White, of Case Western Reserve University in Cleveland, Ohio, demonstrated something similar in a rather gruesome transplant experiment on rhesus monkeys. White was able to transfer the whole head of one monkey onto the body of another. The animals survived for up to eight days after the transfer, could eat normally, and were able to follow movement with their eyes.

White, perhaps mercifully, is now retired, but as recently as 2001 he was still advocating the use of head transplant surgery on humans, to treat patients who were already quadriplegic but who were experiencing organ failure or other multiple complications that would require a number of operations to correct by conventional methods.

Some technologists, such as Robert Freitas, believe that medical nanorobots could help make inroads into the repair of serious amputation trauma—of the sort sustained during a head or brain transplant. Such views are by no means universally held, however.

Another of the key obstacles facing transplant surgeons is tissue rejection—transplant organs being recognized as foreign and attacked by the recipient's immune system. One solution has been to use immunosuppressant drugs. When New Zealander Clint Hallam received the world's first hand transplant in 1998, he was given a fierce cocktail of immunosuppressants. Yet despite this, he experienced serious rejection problems and the hand eventually had to be amputated. Doctors claim, however, that the procedure failed because Hallam was a poor patient who didn't take his drugs regularly. Nearly 30 hand transplant operations have been performed since, with far greater degrees of success—including the first double hand transplant.

One solution to the rejection problem might be to use embryonic stem cells to grow transplant tissue that is an exact genetic match to the patient and so has no chance of being rejected. Stem cells, as we saw in Chapter 14, are mutable cells found in human embryos that have yet to differentiate into particular types of tissue such as blood, brain, or skin. Growing stem cells is controversial because it involves making cloned embryos of the patient. Stem cells can then be extracted from the embryos and injected into damaged areas of tissue, where they will be encouraged to grow into new tissue of the type needed to repair the damage.

This sort of treatment could remove the need for a transplant. Some researchers have suggested that stem cells could be used to grow entire new transplant organs in vitro. Fabre thinks this is unlikely, though. "If you're asking me whether you can put all these things into a bottle and come out with a heart after a month or two, I think that's being wildly optimistic," he says.

Animal Hospital

The body that Solon is assembling in "The Brain of Morbius" is made up of parts from a host of different species. In fact, doctors are already carrying out cross-species transplants—or xenotransplants, as they're known. Since 1975, hundreds of thousands of patients have received replacement heart valves from pigs. And since 1981, valves from cows have also been used.

These valves are not made of living tissue. They have been pickled and thus sterilized prior to transplantation. However, it would be impossible to sterilize a living organ such as a heart or a kidney in this way. And that's why concern over the transfer of diseases from the animal donor to the human patient has become a major safety issue with the xenotransplantation of live organs.

The body has many barriers to prevent the transfer of microorganisms from one individual to another or from one animal to another. But if an animal organ is placed straight into a patient's circulation, all of these safeguards are immediately breached. "Add to that the fact that the patient will be immunosuppressed and you have an ideal environment for a virus within the organ to mutate and become active," says Fabre.

The really big fear isn't that the patient will die from the animal virus, although that's obviously a concern—it's that the disease may mutate within the patient into a form that can then be passed on easily to other humans. It's happened before. AIDS began as a monkey virus that later jumped to human beings, meanwhile avian and so-called swine flu have both mutated in recent years to become transmissible from person to person. "Imagine if AIDS was spread by coughing," says Fabre.

The release of such a deadly infectious disease that could spread with the ease of the common cold would doubtless change our society. "No one would use public transport or go to the cinema. And you'd be careful who you sat next to at work," Fabre adds. "As far as I'm concerned, this is a big risk and effectively knocks the field of xenotransplantation on the head."

So things don't look awfully good for Morbius's brain. In fact, it looks as if it'll be staying in its nutrient tank for some time to come, which can only be good news for the Doctor. In the serial, the Doctor's head was due to house the Time Lord criminal's gray matter and be the crowning glory of Solon's grisly creation.

23

K-9 and Company

> *"Well, that was a piece of cake, eh, K-9?"*
> *"Piece of cake, Master? Radial slice of baked confection . . . coefficient of relevance to Key to Time: zero."*
>
> —The Fourth Doctor and K-9, "The Pirate Planet"

In the Fourth Doctor adventure "Meglos," actor Bill Fraser agreed to appear only on the condition that he would get to kick K-9, the Doctor's robot dog. K-9 was built by Professor Marius of the Centre for Alien Biomorphology on the asteroid K4067. At the end of the serial "The Invisible Enemy," K-9 elects to leave his creator and travel with the Doctor and his female warrior assistant Leela. "I only hope he's Tardis trained," remarks the Professor.

Robo-pooches are now no longer the fictional fancy they were when K-9 made his debut in 1977. In 1999, electronics giant Sony launched the first of its Aibo robot dogs. The artificially intelligent pets can see, hear, touch, and make decisions of their own accord. Their "personality" evolves over a course of months depending on their surroundings and the personality of their owner, and their stereo microphones can recognize around 100 different voice commands. Later models of the Aibo could also play back MP3 music files, receive commands by email, and download Microsoft Outlook appointments from its owner's PC and read them out using text-to-speech software.

The product was discontinued in 2006, but its place has since been taken by other cyber pets—for example, the Ugobe Pleo robot dinosaur. Many experts now believe robotic pets are here to stay. Cybernetics expert Kevin Warwick of the University of Reading in England conducted an experiment in 2000, in which he replaced a family's well-loved dog with a Sony Aibo for a week. "At the end of the week we asked the two kids in the family if they now wanted their original dog back or if they wanted to keep the Aibo," says Warwick. "We were amazed when they said they wanted to keep Aibo."

Aibo-like robots might make ideal low-maintenance companions for the aged. And in our increasingly "money rich, time-poor" society, pets that need no exercise or feeding, and that make no mess, could prove very appealing—especially when they can do so much more than their biological counterparts.

The Doctor himself enjoys playing K-9 at chess, not something the average Labrador would be up to—especially given that K-9 usually wins. In the story "The Androids of Tara," K-9 states that he's been programmed with the moves of all chess championship tournaments since 1866.

In reality, chess computers are already capable of trouncing the best human players. In 1997, IBM's Deep Blue computer famously beat then world champion Garry Kasparov. But scientists say this is no big deal. Humans are generally lousy at chess. We're general-purpose machines that have evolved to perform a wide range of tasks—board games not being one of them. Chess computers, however, are highly specialized machines programmed solely with the rules of chess and given vast computing power with which to implement them. These machines are easy to build, and so it's really no wonder that they can beat us.

True artificial intelligence (AI), of the sort exhibited by K-9, amounts to much more than playing chess and is accordingly very much harder. Independent AI expert Steve Grand, based in Flagstaff, Arizona, uses the simple example of asking a robot to fetch you a glass of water. How does the robot know what water is? How does it know what a glass is in this context? How will it recognize one? How does it recognize the tap when it sees it? How does it coordinate its "hands" to turn the tap on and off at the right moments? Try writing down the precise set of instructions you'd need to program into your robot and you'll soon realize that it's a lot harder than you think, even for this basic task.

Now imagine trying to build a machine that you can hold a rational conversation with. Or one that can feel emotions. Or one that will get your jokes. Building a truly artificially intelligent robot today is an extremely tall order. Grand believes that the big breakthrough we need is to figure out exactly how animals' brains work. He's certain that the operation of the brain is nothing like the operation of a modern electronic computer. But he's optimistic that once we've figured it out, it won't be hard to replicate using technology. How soon? "Any time between a week and a century from now," he says.

The War Machines

When machine intelligence does finally rival our own, how can we expect robots to treat us? Will they try to take over, as has so often been supposed in sci-fi stories, including *Doctor Who* on many occasions? Warwick thinks the answer is a "yes."

He says it will hinge on two key factors: intelligence and control. Humans' rise to dominance on Earth is partly down to our superior intelligence over other species. So if robot intelligence begins to draw level with our own you might understandably worry, although there's no need to, so long as we retain Warwick's second factor—control, that is, the ability to pull out the plug.

You might think that we'd always have the power to switch off a machine that starts to have ideas above its station, and yet there's already a counterexample: the Internet. Practically, we're probably too dependent upon the net to want to shut it down, but even if we did, it comprises so many disconnected elements that shutting them all down would be an enormous and lengthy task and perhaps not even possible over a usefully short timescale. "Even now we're not completely in control of what's going on," says Warwick.

He believes that the first robots with a threatening level of intelligence will probably be developed by the military. K-9 saves the Doctor's life on many occasions, to become the Time Lord's best friend—due in part to the fact that he's armed. His nose laser can kill or stun foes and cut through walls, power cables, or anything else that gets in the way. How realistic is this? Can we really expect machines to be entrusted with the firing of deadly weapons? The alarming truth is that they already are. Cruise missiles—considerably more dangerous than K-9's nose laser and with onboard computers that are far stupider—are already autonomous, making many of their own guidance decisions. And if that shocks you, the American military is already using UCAVs, uncrewed combat air vehicles—remotely operated fighter planes—in Afghanistan and Pakistan. As of the spring of 2009, the US Air Force had over 220 robot fighter drones in operation and it's estimated that they had already killed some 100 Taliban militants.

"I find this quite worrying," says Warwick, "particularly the concept of the complete removal of any human overriding control." At the time of writing, American robot planes are remote controlled by human pilots on the ground and not yet fully autonomous.

The *Doctor Who* writers have penned some frightening depictions of Warwick's worst-case scenario. In "The Robots of Death," humanoid robots are subverted to kill humans at the bidding of the evil villain Taren Capel. And in 1966's "The War Machines," the supercomputer WOTAN (Will Operating Thought ANalogue) becomes self-aware and constructs an army of War Machine robots to wipe out all of humanity. As well as superior intelligence, WOTAN also develops human-like psychological powers—for example, it can hypnotize victims, even over the telephone.

Warwick thinks that this is perfectly feasible, arguing that a superior electronic brain would almost certainly end up knowing more about our organic brains than we do—and so would be in a strong position to manipulate them in this way.

So we can't pull the plug out, and we'd best not try negotiating with them on the phone—what exactly are we going to do when the robots rise up against us? In "Robot," Tom Baker's first foray as the Fourth Doctor, the experimental robot K1 goes on the rampage, killing its creator and threatening to destroy the Earth. The Doctor eventually defeats the mechanical monster using a "metal virus" originally developed to rid the world of metallic waste by rapidly corroding it away. How could that work?

"The notion of an organism that eats metal isn't totally farfetched," says Fred Glasser, a chemist at the University of Aberdeen, in Scotland. He says that many organisms eat via processes involving *oxidation* and *reduction*—chemist-speak for corrosion, or rusting. He imagines microorganisms that are able to trigger a chemical reaction with a metal, causing it to give off energy, on which the organisms can feed, while rusting the metal away in the process. Of course, just because it's an organism doing the dirty work, it doesn't mean that we can necessarily call it a virus. As we saw in Chapter 10, viruses work by hijacking the DNA-copying machinery in the nuclei of cells that they infect—and using this machinery to replicate themselves. Clearly, this won't be possible here, simply because metal has no DNA. Then again, nobody's suggesting that computer viruses have anything to do with DNA either.

If we're prepared to relax the condition that the "metal virus" be organic, then there are certain chemicals that can do a pretty good job of reducing metals to dust. These penetrate into the metal's grain boundaries (the areas where different crystals in the metal join together), weakening them enormously. A good example is aluminum. When this metal is exposed to the element gallium it becomes spectacularly weakened—to the point where it almost literally falls apart. So don't take any gallium with you next time you travel on a plane.

Still, even with metal viruses to arm ourselves against the robots, I don't really want to spend the future fighting a Cyber War. And I'm sure you don't either. Just as well, then, that not all robotics researchers share Kevin Warwick's grim vision.

Steve Grand, for one, is rather more optimistic. He points out that truly intelligent machines might realize the futility of war and have the sense to be pacifists. Indeed, perhaps this is why the Doctor trusts K-9 with a deadly weapon even though he refuses to bear arms himself.

K-9 is a good example of how we might want our future relationship with robots to be. As well as guarding the Doctor and beating him at chess, K-9 is a valued assistant, able to sense danger, solve problems—and even read him extracts from *The Tale of Peter Rabbit* (see "The Creature from the Pit"). How does that compare with real robot assistants today?

The Android Invasion

You can already buy robot vacuum cleaners to clean your rugs, robot lawn mowers to cut your grass, and even robot mops to wash your kitchen floor. And sophisticated robot assistants are now used in dangerous professions such as bomb disposal and fire-fighting.

Robotics expert Noel Sharkey, of the UK's University of Sheffield, says that the time is even near when human office workers will have assistants in the form of talking humanoid robots fixed to their desks.

There's considerable motivation for developing robot assistants that resemble humans, similar to the androids that appear from time to time in the show—such as Kamelion, which the Fifth Doctor gains as an assistant in "The King's Demons," and the Anne Droid from the Ninth Doctor episode "Bad Wolf."

"In a human environment, as would be the case with domestic helper robots, where you want a robot that can go places where humans can go, then there is certainly a case for humanoid robots," says Warwick.

Robots will also play a key role in space, an area close to the hearts of many *Doctor Who* fans. Researchers at NASA's Johnson Space Center (JSC) at Houston in Texas are working on Robonaut—a humanoid robot astronaut that can work with humans on space walks or even perform on its own when the risks are too great for people. Under current plans, Robonaut will be controlled remotely by a human operator. But as new breakthroughs in AI are made, machines like this will become increasingly autonomous.

Will they ever reach the level of sophistication of K-9? In other words, will we see wheeled mobile robots assisting travelers on the surfaces of other planets? That's what NASA is hoping. In 2006, Robonaut was fitted to a four-wheeled chassis to create a future concept to assist human astronauts with reconnaissance, transporting equipment, sample collection, and other fieldwork. The idea is to produce a wheeled autonomous robot that can recognize spoken commands and hand gestures and can determine from body movements what an astronaut is doing so it can offer help accordingly.

It's not quite up there with K-9 yet, but Robonaut has stolen one crucial march on its fictional counterpart. The radio-controlled K-9 prop used in the filming of *Doctor Who* was often let down by Dalek syndrome—its tiny wheels and low chassis struggling over rough outdoor terrain. The solution usually involved the production team laying down planks and pulling K-9 along with a concealed piece of string. Robonaut may have some way to go before it can match K-9's fluent chatter and high-tech capabilities, but it does have one thing going for it—seriously big wheels.

24

Psychic Paper

"Really, Doctor—you'll be consulting the entrails of a sheep next."

—Brigadier Lethbridge-Stewart, "The Time Monster"

It's the gatecrasher's dream come true: an invite that can get them past the bouncers at just about any party they feel like going to. The Doctor calls it psychic paper. When flashed before someone's eyes it shows them exactly what they expect to see or what the owner of the paper wants them to see—whether that's your invitation to come and witness the destruction of the Earth (as in the Ninth Doctor episode "The End of the World") or what you want a potential lover to know about you (as in the great piece of dialogue between Rose and Captain Jack in "The Empty Child").

If a device like this were to work, it would need to be able to measure brain activity, translate the electrical data it measured into a palpable image, and then display this image on a compact screen.

In April 2005, *New Scientist* magazine reported on a mind-reading machine that could do some of these things, built by researchers from Japan and the U.S. Yukiyasu Kamitani at ATR Computational Neuroscience Laboratories in Kyoto and Frank

Tong of Princeton University used a functional magnetic resonance imaging (fMRI) scanner to deduce what a subject is looking at, purely by analyzing their brain activity.

MRI is a scanning technique that coaxes radiation from hydrogen atoms inside the body's water molecules. It works by passing a strong magnetic field through the body. This causes the electrically charged nuclei of the hydrogen atoms to snap into alignment with the field. Next, a pulse of radio waves is passed through the area to be scanned. This briefly flips some of the hydrogen atoms in the area out of parallel with the field. As the atoms flip back they give off radio waves, but different types of tissue flip back at different rates, and this is what reveals the structure of the body within. The magnetic fields used are intense—up to 3 Tesla, or 100,000 times the strength of Earth's magnetic field.

To examine brain functions, fMRIs look in particular at blood flow, which shows the parts of the brain that are using the most oxygen—these are the areas that are the most active.

The researchers showed four volunteers patterns of parallel lines that could be in one of eight different configurations. Looking at each pattern of lines created a different pattern of brain activity in the volunteers, which the researchers were able to detect with the fMRI scanner. Interestingly, they were even able to tell which pattern had been presented when it was flashed up too fast for the volunteer to recognize it consciously.

Clearly, this kind of technology has some way to go before it could be used in psychic paper, or anything even remotely similar. But it demonstrates the feasibility of measuring what someone is thinking using an external device. Now all the *Doctor Who* writers need to do is figure out how to condense an fMRI machine—typically the size of a room—down into something that can fit inside a sheet of paper. The technologists, it would seem, have their work cut out for them.

Think Small

Cropping up in Chapters 12, 21 and 22, nanotechnology—engineering on scales of a billionth of a meter—is something of a magic wand when it comes to explaining the various technologies of *Doctor Who*. And, yes, you've guessed it—it's one possible future way to make psychic paper work as well. "By positioning a nanorobot in each of the brain's 10 billion cells, it becomes possible to directly read and transmit real-time brain states," says Robert Freitas of the Institute for Molecular Manufacturing in California. "However, I'm not so sure the response can happen as quickly as hoped, unless everyone has lots of nanorobots already embedded in them."

The difficulty here is getting the nanorobots into the subject's brain over a timescale that's short enough to be useful. They would have to enter the body through one of the various orifices and from there make their way into the bloodstream, swim to the brain, and deploy themselves. All this would very probably take longer than it would for the doorman at a party to send the Doctor packing.

So does this mean that psychic paper is sunk for the time being? Not necessarily. Parapsychologist Chris French of the Anomalistic Psychology Research Unit at Goldsmith's, University of London, thinks that the properties of psychic paper bear uncanny similarities to a notion in psychology research known as "top-down" processing.

This is the tendency of people to see what they're told to see—without any help from psychic paper or other props. "When we perceive the world, we've got two sources of information—on the one hand we've got the raw sensory input, the information coming through our eyes, our ears, and our other senses," says French. "But in order to make sense of that you've also got the influence of top-down processes—your beliefs, your expectations, and your knowledge about the world."

Such is the power of top-down processing that if someone tells you what you're supposed to see or hear when you look at or listen to something, this can seriously skew your perception of it.

A classic example is the backward message supposed to exist in the Led Zeppelin song "Stairway to Heaven." This is one of the most stunning examples of top-down processing that you're ever likely to meet. It's in the verse of the song that starts: "If there's a bustle in your hedgerow . . ." Get on the Internet and go to: http://jeffmilner.com/backmasking.htm, and then click "Stairway to Heaven," top right. You'll be presented with the normal lyrics, links that let you play the verse both forward and backward, and a link to display the supposed backward lyrics. First of all, play the verse backward without reading what the backward lyrics are meant to be. You might pick up one or two words, but in general it's gobbledygook. Now read the backward lyrics and play the sound file again.

"You can now hear the message as clear as a bell. You don't know how you missed it the first time," says French. "The point is that until you're told what to hear, you don't hear it."

This message wasn't placed here by the members of Led Zeppelin—and not even by Satan. In fact, there's no message here at all; it's just a matter of statistics. If you play thousands of hours of rock music (or any other music with lyrics) backward, looking for messages, then sooner or later you're going to find some. "These people really should get out more," says French.

On the same website you can find a whole raft of other songs with backward messages that are equally inscrutable—until, again, you're told what it is that you're meant to hear. Tricksters and conjurers such as Derren Brown and Penn & Teller exploit top-down processing all the time to make us see and think what they want us to. And for some reason we seem to be endlessly impressed by our own gullibility.

So do you really need the Doctor's psychic paper to talk your way into that party? Probably not. The truth is that a wily mind and the power of suggestion are very often enough.

25

Space-flight

"We're not on a yacht; we're on a ship . . . a spaceship!"

—The Fifth Doctor, "Enlightenment"

It would be a rather dull state of affairs if the Doctor was the only space traveler in the Who universe. Far from it, though. His galaxy is chockfull of alien races plying the space highways in all manner of interplanetary and interstellar vehicles. It makes the modern-day efforts of organizations like NASA and the European Space Agency look decidedly feeble. But is it an accurate picture of the future? What will human spaceflight really be like hundreds or maybe thousands of years from now?

So far, we humans have only dabbled in interplanetary spaceflight. We've sent crewed rockets into orbit around the Earth and to land on the Moon. And we've sent robotic space-probes farther afield to investigate other planets and their moons.

At the moment, interplanetary travel is a slow and costly business. Chemical propellant is burned and the hot exhaust then ejected from the back of the rocket engine at high speed to propel the spacecraft forward. Rockets work because of a principle from physics called conservation of momentum. It says that motion of matter in one direction must be balanced by motion of matter in the opposite direction. It's the same reason that a rifle kicks back against your shoulder when you fire it—the rapid motion of the low-mass bullet is balanced by slower motion of the larger mass gun back toward you. In the case of a rocket, a small, fast-moving mass of fuel is ejected, causing the larger-mass rocket to move forward. But just to send a 10,000-kilogram probe from Earth orbit to Mars this way could take as much as 200,000 kilograms (more than 50,000 gallons) of chemical fuel—a vast amount.

So what are the alternatives? In the 1983 Fifth Doctor serial "Enlightenment," the Doctor finds himself aboard a spacecraft with a difference. It's more like a sailing ship in space than a rocket. Oddly enough, engineers are investigating exactly this kind of spacecraft as one way to travel around the Solar System more efficiently. Called solar sails, these vessels use a large, reflective silver sheet to catch the light streaming out from the Sun. When a photon, a particle of light, hits the sail it bounces back in the opposite direction. Conservation of momentum then causes the sail to move forward. Steering is handled by a number of smaller vanes to tilt the sail with respect to the light from the Sun, just as a yacht's sail is tilted to steer it in an ocean breeze.

The acceleration exerted on a solar sail by the radiation pressure from the Sun is very modest. Imagine how fast a stone falls out of your hand when you drop it on Earth. A typical solar sail in the inner Solar System (where the Sun's light is strongest) would accelerate at just one ten-thousandth of the rate the stone falls at, gaining speed at just 1 millimeter per second every second. That means that to accelerate from 0 to 100 kilometers per hour would take several hours. Reaching interplanetary speeds takes months to years.

But the wait could be worth it. "By making a close pass by the Sun, a sail can accelerate up to high cruise speeds," says Colin McInnes, who researches solar sails at the University of Strathclyde, Scotland. Missions to the very outer edge of the Solar System—to distances 200 times farther than the Earth is from the Sun (200 astronomical units, or AU)—could take just 25 years. Compare that with the *Voyager 2* probe, which as of September 2009 had been in space for 32 years and had reached a distance of just 90 AU.

The other big advantage is that once a solar sail is in space it uses no fuel—at least not while it's in the inner Solar System. Once the sail reaches beyond the orbit of Jupiter, sunlight becomes too weak. However, engineers say that one option to accelerate it on from there would be to direct powerful lasers onto the sail from Earth. "Laser sails are in principle possible with a cruise speed of around 0.1 lightspeed," says McInnes. "But they require huge laser power."

In 2004, the Japanese space agency, ISAS, successfully deployed the world's first solar sails in space. Two prototype sails, each just 7.5 micrometers thick, were unfurled in Earth orbit. And there have been tests of much larger sails on the ground. In April 2005, NASA's Marshall Space Flight Center in Huntsville, Alabama, began testing two 20-meter solar sail designs that they say could be used as the propulsion systems for future deep-space missions.

To the Stars

What about interstellar space travel? It doesn't take a rocket scientist to work out that unless we make some fairly radical new discoveries in physics, this is going to be a painfully slow process. The nearest stars are light-years away. Trying to send ordinary space rockets this kind of distance would burn up more fuel than there is matter in the entire visible Universe. And even a super-efficient solar sail, goaded along by a powerful laser beam from Earth, would be hard put to make the journey in under 40 years.

For the denizens of the *Doctor Who* universe, skipping around the galaxy seems to be a matter of course. So what do they know that our scientists are missing? In "Mawdryn Undead," the Tar-

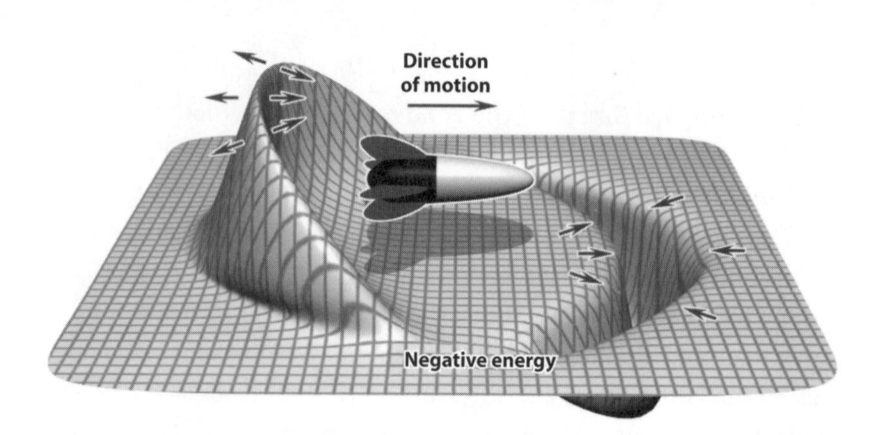

FIGURE 13. In Miguel Alcubierre's concept of "warp drive," negative energy matter placed around a spacecraft causes the space in front of the craft to shrink and the space behind it to expand at the same rate, sweeping the craft to its destination.

dis collides with the "warp field" of a passing starship and is forced to materialize on board. Similar space-warping propulsion systems are mentioned in the adventures "Nightmare of Eden" and "Warrior's Gate."

In 1994, physicist Miguel Alcubierre of Cardiff University in Wales calculated how such "warp drive" propulsion systems might actually work (see figure 13).

Conventionally, spacecraft are constrained to travel at less than the speed of light. Albert Einstein's special theory of relativity posited lightspeed—300,000 kilometers per second—as the maximum speed at which any object can move through space. Alcubierre's trick was to forget trying to move the craft through space but instead to bend and stretch the space around it to form a kind of wave that sweeps the spacecraft along to its destination.

He used general relativity (Einstein's theory of gravity—see Chapter 3) to show how this could be done if the craft is surrounded by a bubble of so-called exotic matter—the same negative-energy material that we met in Chapters 2 and 3. If the exotic matter was arranged around the spacecraft in just the right way it could cause the space in front of the craft to shrink rapidly, while the space behind it expanded at exactly the same rate, carrying the piece of space containing the craft along to its destination arbitrarily fast.

Since it was first proposed, Alcubierre's warp drive idea has attracted scathing criticism from some physicists, who argued that the amount of exotic matter that it would require is unfeasibly large. Some even argued that it would take a negative mass equal to some—6×10^{62} kilograms (a number 6 with 62 zeroes after it), or 10 billion times the mass of the entire visible Universe, to do what Alcubierre was suggesting.

Modern estimates look slightly more optimistic—thanks to exactly the same piece of research that could make the Tardis bigger inside than out, which we looked at in Chapter 2. The research was carried out in 1999 by Chris Van Den Broeck, then at the Catholic University in Leuven, Belgium. He showed how you could make a faster-than-light spacecraft effectively much smaller by putting it inside a region of space that's bigger inside than out. That means you can get away with a much smaller bubble of exotic matter to make your warp drive work—even after taking into account the extra exotic matter that you need to create this Tardis-like space in the first place.

"The Tardis allows you to make the warp bubble smaller so that it requires less energy," says Van Den Broeck. Indeed, his result reduces the amount of exotic matter required by a colossal factor of 10^{33}, to -6×10^{29} kilograms, about a third of the mass of the Sun. (Exotic matter can have negative pressure and mass. As we saw in Chapter 2, it's allowed to exist because the energy—or, equivalently, the mass—of empty space is non-zero thanks to quantum particles constantly appearing and disappearing. Anything with less than this non-zero *vacuum energy* effectively has a mass that is negative. The Casimir effect, also discussed in Chapter 2, is one way to create exotic matter.)

Even so, Van Den Broeck is still doubtful that it will ever be possible to build a warp drive for real. In more recent research, he has shown that Alcubierre's scheme suffers from a far more crippling flaw. He says that in order for a pilot to control the warp field around his spacecraft he would have to send signals between the craft and the warp field generators at faster than the speed of light.

But hang on—isn't going faster than light what warp drive is all about? Yes and no. The warp drive idea works by deforming space so that the spacecraft moves faster than light when you step back and look from a distance. Physicists say that it breaks the speed of light globally, and it seems that this is allowed under certain circumstances. But remember that locally the spacecraft isn't going anywhere. It's just sitting still in a piece of space, and it's this piece of space that's getting whisked along at warp speed.

Van Den Broeck is claiming that to make warp drive work you also need to break the local speed of light inside this piece of space. "And this is absolutely not allowed by anything we know of," he says.

Radiation Sickness

If all this discussion about the difficulties in creating warp drive has put something of a damper on your hopes for interstellar space travel, then Frank Close, a physicist at the University of Oxford, has some even worse news in store for you. In 2005, Close headed a panel of three scientists commissioned by the UK's Royal Astronomical Society to review the scientific case for sending human beings into space. The panel found that propulsion is the least of our worries. A far more pressing problem that will need solving before we can even contemplate the crewed exploration of the Solar System, let alone the rest of the galaxy, is how to deal with the vast amount of lethal radiation streaming through space.

"Radiation in space probably makes a human mission to Mars just about okay," says Close. "But beyond that it's deadly—going as far as Jupiter would give an accumulated radiation dose akin to what you would get in cancer therapy." Trying to go to the stars would be suicide.

The most hazardous radiation in space takes the form of what are known as "galactic cosmic rays"—which aren't rays at all, but positively charged atomic nuclei that have been accelerated to huge speeds by high-energy cosmic phenomena like supernovae, the explosions marking the deaths of massive stars. It's been estimated that a single cosmic-ray particle can pack as much punch as a fast tennis serve. If that kind of energy crashes into the DNA inside a human cell, it can lead to potentially fatal mutations and cancer, and damage the body's machinery for making new blood cells.

Solar flare particles from the Sun have also been identified as a radiation hazard to astronauts, though all but the most severe of these can be staved off by cladding spacecraft in a layer of lead shielding. And this is exactly what robotic space-probes use to guard their sensitive electronics as they journey through interplanetary space.

But lead shielding isn't much use when you want to protect your human crew from galactic cosmic rays. All that happens is that an incoming cosmic ray particle smashes into a lead nucleus in the shield so fast that a shower of high-speed debris—neutrons, electrons, and electromagnetic radiation such as x-rays—cascades out the other side. And this "secondary radiation" is almost as deadly as the cosmic ray itself.

Ironically, despite presenting its own radiation hazard, the Sun offers some degree of protection against galactic cosmic rays. As we've seen, the solar wind of charged particles constantly streaming out from the Sun forms a buffer that reduces the intensity of the rays as they enter the Solar System. However, the degree of protection afforded by the Sun lessens substantially as you travel farther away from it.

It's been suggested by some physicists that because cosmic rays are electrically charged they could be deflected by fitting spacecraft with giant electromagnets. This idea has been greeted with skepticism in the past, with critics pointing out that until such an electromagnet can be made sufficiently "non-giant" to fit in a space rocket, it's not going to get off the ground, let alone into space.

But in 2008, a team of scientists led by Ruth Bamford at the Rutherford Appleton Laboratory, in England, figured out how it might be done. "It was thought that the magnetic bubble surrounding a spacecraft had to be around 20 kilometers across, making the magnet on the spacecraft massive and requiring megawatts of power," she says. "What we have now found—in theory, computer simulation and laboratory experiment—is that a magnetic bubble just 100 meters across would be sufficient to protect the spacecraft."

And this would mean that the electromagnet could be of practical mass and dimensions to be carried into space aboard a present-day spacecraft. There's still much research to be done before human astronauts can benefit from electromagnetic shields such as this, but it's the start of a solution that could one day enable humans to set sail across the galaxy.

26

Space Stations and Moonbases

"This isn't actually the Fourth Great and Bountiful Human Empire. It's a place where humans happen to live . . . Sorry . . ."

—The Editor explains the purpose of Satellite 5,
"The Long Game"

Artificial habitats in space pop up rather often in *Doctor Who*—from the Wheel in Space encountered by the Second Doctor to Platform One, where the Ninth Doctor and his assistant Rose witness the demise of the Earth.

As humans expand their domain outward from Earth, the number of bases in space and on other planets can only increase. Our first ever space station was the Russian *Salyut 1*, launched in 1971. It was a cylindrical craft just 20 meters long and 4 meters across, capable of carrying a three-person crew. There were a further six *Salyuts*, followed by the star of Soviet long-duration spaceflight, the multi-moduled *Mir*, which stayed aloft for a staggering 15 years. *Mir* wasn't without its share of problems, though. In 1997, an on-board fire, failing oxygen generators, and a collision with a robotic supply vessel left crew members fearing for their lives. *Mir* was eventually de-orbited in 2001.

The first U.S. space station was *Skylab*. It was built inside the third stage of a Saturn IB rocket—since it wasn't going to the Moon, the fuel that would normally have filled that stage wasn't needed. *Skylab* was launched in 1973 and stayed in orbit for 6 years. "Although not that big on the outside, it was very roomy on the inside," says Piers Bizony, author of *Island in the Sky: Building the International Space Station.* "It is still the largest spacecraft in history to be launched in one go."

Skylab eventually fell back to Earth when increased solar activity heated the outer layer of the Earth's atmosphere, causing it to expand and drag on the station, degrading its orbit. *Skylab* had been abandoned as of February 1974; however, one Australian cow was killed on the ground when it crashed to Earth on July 11, 1979.

Now, the only human space outpost in service is the International Space Station (ISS). It's a collaboration between six world space agencies (the U.S., Russia, Canada, Japan, Brazil, and Europe). The first module was launched in November 1998, and the ISS has been permanently crewed since November 2000. Construction of the station was severely hampered by the grounding of NASA's space shuttle fleet following the *Columbia* space shuttle disaster in 2003. And with the shuttle's retirement now looming, several elements of the ISS have been cancelled. Nevertheless, as of March 2009, 81 percent of the remaining elements in the ISS plan had been installed, with the last—the Russian Multipurpose Laboratory Module—due to be launched in December 2011.

Space Hotels

The next space station could be just around the corner. And, rather than being another dull science lab, this could be something much more fun: a hotel. In July 2006, Bigelow Aerospace of Las Vegas launched the first of two experimental *Genesis* inflatable space station modules. The modules, made from super-high-strength fabrics, were boosted into space in a collapsed state and then inflated. Genesis is a cut-down model of the full-scale hotel module and will be used for rigorous on-orbit testing before the real hotel is launched, which Bigelow hopes will happen in the next few years. The second module was launched in June 2007, and both are still in orbit.

The project, the brainchild of Robert Bigelow (owner of the Budget Suites of America hotel chain), couldn't have come at a better time. In 2005, Californian aeronautical engineer Burt Rutan claimed the X-Prize—the award for the first private crewed space launch. Shortly after, five scaled up versions of Rutan's craft were ordered by tycoon Richard Branson for his Virgin Galactic fleet of commercial spacecraft, which is expected to begin flights in 2011. Space tourism is promising to be a booming industry in the very near future.

Indeed, in the future explored by *Doctor Who,* it certainly seems to have caught on. We meet space passenger liners on several occasions, such as in "Nightmare of Eden" and "Terror of the Vervoids," while in "The Leisure Hive" the Doctor visits a pleasure complex on the planet Argolis.

Bases on other planets and moons are something NASA is now looking at. In January 2004, President George W. Bush announced plans for NASA to return to the Earth's Moon by 2020, establish permanent crewed bases there, and then move on to Mars. In September 2005, then NASA chief Michael Griffin announced how the agency plans to achieve Bush's goal. The grand scheme, spread over 13 years and costing an estimated $104 billion, will begin with robot scout missions to the Moon by 2011 to investigate possible crewed landing sites. The first human landings will take place by 2020, with astronauts remaining on the lunar surface for roughly a week at a time.

Astronauts will use the Moon as a proving ground to develop the skills needed to live on the surface of Mars. This includes learning to "live off the land," for example splitting water ice found on the Moon into hydrogen and oxygen for fuel and breathing. For this reason, a likely landing site for the first

Moon missions is the lunar south pole, where several robotic space-probes have already found evidence for the existence of water ice in permanently shaded craters.

Griffin didn't set a date for the establishment of a lunar base in his announcement, but Bush's presidential directive certainly called for one. As well as preparing astronauts for living on Mars and other destinations even farther afield, an eventual base on the lunar far side would also be ideal for astronomy. Radio telescopes there would be shielded from the constant hiss of electromagnetic noise from human industry that plagues radio astronomy on Earth.

Red Is the Color

Back during the heyday of the *Apollo* era, space scientists discussed the idea of sending crews on to Mars as early as 1986. Now it looks like the earliest this can happen is in the 2030 time-frame, or more likely beyond. NASA's plans for sending humans to Mars as part of Bush's directive haven't yet been announced. However, aerospace engineer Robert Zubrin—president of the Mars Society, a group campaigning for the human exploration of Mars—has long advocated his "Mars Direct" plan as the quickest and most efficient way to send a human crew to the red planet.

The plan involves first robotically piloting an Earth-return vehicle (ERV) ahead to Mars. This is a spacecraft that can carry humans on the return journey from Mars to Earth. The imagined scenario would go like this: The ERV has on board a small tank of hydrogen and a nuclear reactor. Once on the surface, power from the reactor is used to combine the hydrogen with carbon dioxide from Mars's atmosphere to make large amounts of methane and oxygen—which can be mixed to make a potent rocket fuel. Mars is at its closest to Earth every 26 months. So 26 months after the launch of the ERV, a second spacecraft sets off from Earth—the habitat module, this time carrying a human crew of four. Prior to launch, ground controllers first radio the

ERV on Mars to check that it has manufactured sufficient fuel for the return flight to Earth—so there's no chance of the crew's being stranded on Mars by the ERV failing to fuel itself. The habitat module takes approximately 6 months to get to Mars, and then lands in the vicinity of the ERV.

The crew would spend about 18 months on the surface of the red planet before using the ERV to return home, leaving the habitat module behind for the possible use of future visitors. The cost of Mars Direct has been estimated at around $30 to $35 billion (in 2004 dollars), about a third of the conjectured cost of NASA's Moon plan and quite within the realms of what the agency can afford.

On both legs of the journey, the upper stage of the booster rocket is retained and attached to the crew module by a long tether. The booster acts as a counterweight so it and the crew module can then revolve around their common centre of gravity like a bola, providing the crew with artificial gravity. The craft spins at just the right rate so that the centrifugal force it places on the crew has the same strength as gravity on Earth.

Artificial gravity is important on a long-duration space mission. Without the force of gravity constantly pulling on an astronaut's body, the muscles begin to waste away at an alarming rate. On the International Space Station, there's no artificial gravity—instead, the astronauts exercise for 2 hours a day to preserve their muscle mass. But that's not the only health hazard of low gravity. It depletes bone calcium, weakens the immune system, and even causes excess flatulence. And that's not to mention the general malaise of space sickness—vomiting, loss of balance, headaches, and lethargy—that affects 45 percent of space travelers and can last anything up to 72 hours while the body adapts to its new weightless environment.

So it's no surprise that artificial gravity is a feature of most space stations in Doctor Who. In particular, the wheel from "The Wheel in Space"—like Zubrin's Mars Direct spacecraft—also generates artificial gravity by rotating. And the wheel is resupplied by cargo vessels known as "silver carriers," remarkably similar to the uncrewed Russian *Progress* craft currently used to resupply the ISS.

It's unclear whether President Obama will continue Bush's vision for space exploration. And even if he does, it's unclear where NASA will take the crewed exploration of space to next, once it's done the Moon and Mars. The creators of *Doctor Who* certainly haven't held back, conjuring human outposts in the chilly organic murk of Saturn's moon Titan and even on the lonely, rocky surface of an asteroid (see Fourth Doctor serial "The Invisible Enemy"). Building a base on an asteroid isn't such a crazy idea and was seriously suggested by physicist Freeman Dyson (see Chapter 19).

Even a Titan base isn't impossible. Assuming the cost of transporting the necessary infrastructure to the outer Solar System and then keeping it resupplied could be met, the physical challenges of living there might not differ greatly to living aboard the ISS today. Once you've cracked the problem of keeping what's inside in and what's outside out in space, doing it elsewhere may not be such a huge problem. The challenge, as we saw in the last chapter, will be transporting your crew all the way out to Titan without their dying of radiation sickness en route.

Colony in Space

It's been predicted that, in the far future, humans will reside in giant rotating cylinders in space. In 1974, space visionary Gerard K. O'Neill, a physicist at Princeton University, published an article in the magazine *Physics Today* outlining how humans could leave Earth and live on the insides of great cylinders in space, each 3 kilometers in radius and 30 kilometers long. The cylinders, which have since become known as O'Neill colonies, would rotate on their axes to create artificial gravity. Three vast "window stripes" running the length of each cylinder would allow sunlight in, and a clever arrangement of mirrors would reflect the light down onto the land areas. A thick shell of steel and glass, and the vast quantity of air within the cylinder, would serve to block out cosmic radiation.

O'Neill died in 1992. But he was freshman advisor to Seth Shostak, now a senior astronomer at the Search for Extraterrestrial Intelligence Institute in California, who remembers him well. "O'Neill was predicting that by the 1990s, there would be tens of millions of people living in these rotating aluminum habitats, which as you may have noticed did not occur."

O'Neill's timescales were clearly rather optimistic, but Shostak says that his physics was sound, and he's convinced that O'Neill's dream will one day become reality. The cylinders could be located in Earth orbit or farther away, on the other side of the Solar System—just like Nerva Beacon, the deep-space outpost in the Fourth Doctor adventure "Revenge of the Cybermen."

The only real show-stopper for O'Neill colonies is economics. The explosion in the commercial space launch industry, which we're now just beginning to see, will almost certainly bring about a significant fall in the cost of boosting hardware into orbit. But at the time of writing, launching all but the most basic of structures is prohibitively expensive. "For the moment it's still cheaper to build in Clapham Junction [an area of central London] than it is to build in space," says Shostak. "Even though it may not be quite as desirable."

27

Bombs, Bullets, and Death Rays

"Have you thought up some clever plan, Doctor?"
"Yes, Jamie, I believe I have."
"What are you going to do?"
"Bung a rock at it."

—Jamie and the Second Doctor, "The Abominable Snowman"

The Doctor (with a few exceptions throughout the show's history) refuses to take up arms against his enemies, preferring to use reason and cunning instead. But not all of his adversaries, nor indeed his allies, are quite so cultivated. In fact, many seem perfectly happy to knock lumps out of one another with a frightening array of weaponry, from death rays to sonic disruptors to fission grenades.

In the real world, weapons research is one area where human beings seem to need little encouragement. The feeling of security that having a gun in one's glove-box brings (even though statistics show that the person most likely to get shot with it is the owner) seems to apply at a national as well as a personal level. And when fear and insecurity aren't there to motivate us, there's usually someone somewhere in the midst of a conflict more than willing to pay for the necessary hardware.

Naturally, alien weaponry is likely to be much more advanced than anything that's been developed on Earth so far. But what direction is weapons research heading in? Might the military creations of the *Doctor Who* writers ever materialize on a battlefield—terrestrial or otherwise?

Weapon of Choice

Perhaps the most commonly encountered alien firearms are energy weapons—devices that project a high-energy beam of heat, light, sound, or other form of radiation in order to inflict damage from a distance. Similar devices are being developed in military labs around the world. Scientists have calculated that a one-second pulse from a 1-million-watt laser striking a target would be roughly equivalent to the blast from about 200 grams of high explosive, the damage caused as material in the target is explosively evaporated.

The major difficulty is in making a 1-million-watt laser. That kind of power consumption is, frankly, huge. And generating it as part of a weapon system that can be moved around on a battlefield is a challenge, to put it mildly.

Michael Wartell of Indiana University–Purdue University Fort Wayne, who serves on the U.S. Army Science Board, thinks that limitations in power-generation technology make it difficult to construct a mobile battlefield laser today. "They're certainly not mobile at the present time," he says. "However, there are fixed positions where you can build the gigantic power systems that are required and the large mirror systems and lenses and other apparatus."

That said, Wartell imagines that in the next few years units might be fielded in which the necessary generators and other components are sufficiently compact to be loaded onto the back of a truck, giving the military heavy laser weapons that have a similar degree of maneuverability to a tank.

These weapons would initially be used primarily against stationary targets such as parked vehicles, artillery pieces, installations, and unexploded munitions. The reason for this is that the energy weapons of the immediate future won't be powerful enough to give you the same instant hit that you get with a projectile. "You have to have a dwell time of the laser on the object, typically of a few seconds. And that's the difficult problem with using a laser—tracking and aiming," says Wartell. If you can train a laser on the jet engine of an aircraft or an incoming missile for long enough, then it's possible to destroy it, but the problem is in training the laser on such a fast-moving target for any duration. "I'd argue that's why we're still shooting bullets at things," he says.

Given all the problems that energy weapons seem to throw up, you might wonder if there's really a case for using them. Wartell thinks that in the long run there is. "The first issue is the psychological one. When you think about death rays, that strikes fear into the hearts of everyone," he says. "And you can replenish the source. You have to stick a bullet in a gun, but once you have reliable power on a laser, your ability to shoot is almost limitless."

It's still going to be a very long time until combat soldiers will be drawing a bead on one another with laser rifles and pistols. Daleks and Cybermen take note. However, low-power lasers have been used to great effect in sighting devices on handheld weapons (the ominous red dot) and as range-finders. Another handheld application is in the field of non-lethal weapons. "Our Threat Assessment Laser Illuminator has a range of 350-plus meters and is capable of disorienting and distracting aggressors, as well as stopping vehicles at long range by filling the windshield with a covering green light," says Pete Bitar, of Xtreme Alternative Defense Systems (XADS), a defense technology firm based in Anderson, Indiana. A cut-down version of the pistol-like device was deployed in Iraq in January 2005.

Non-lethal weapons such as this that interfere with the target's visual system are legally required to be of low enough power not to cause blindness (and the XADS system complies with this requirement). It stems from a Geneva convention prohibition on weapons designed purely to maim.

Other researchers have come up with designs for non-lethal energy weapons that can physically stun their targets. In November 1997, *New Scientist* magazine reported on the work of inventor Hans Eric Herr. Based in San Diego, California, he came up with a beam-weapon equivalent of the Taser. Tasers are stun weapons that deliver an electric shock. They consist of a handheld device that shoots two metal darts into the target. Trailing the darts are wires, down which is unloaded an electric shock of around 50,000 volts, causing every muscle in the victim's body to contract, incapacitating them. The idea is that the weapons don't kill because, although the voltage is high, the current they deliver (that is, the number of amps) is still low. (However, according to the American Civil Liberties Union, as many as 148 people in Canada and the United States have in fact been killed by Tasers since they were introduced in 1999, bringing their non-lethal status under serious scrutiny.)

Herr's energy Taser works without the wires, by generating an intense beam of ultraviolet light, which creates a corridor of "ionized" air between the gun and the target. Ionization is the process of separating one or more negatively charged electrons from their atoms. Air that has been ionized is electrically conductive and so a high-voltage Taser shock can then be sent down it. Whereas a dart-and-wire Taser has a range of just a few meters, Herr's design can reach targets up to 100 meters away. And while a Taser is one shot only, Herr's weapon can be recharged and fired repeatedly. The drawback is that when first proposed in 1997 it was the size of a kitchen table. Herr has since sold the patent for the device to HSV Technologies, a defense company allied to XADS.

Stun guns have often featured in *Doctor Who*. For example, in the Third Doctor serial "Frontier in Space," the mercenary Ogrons used energy weapons with a stun setting for subduing victims without killing them. And the Cybermen used a stun weapon—a blinding flash of light emanating from their headsets—on the Doctor's assistant Polly in "The Tenth Planet."

Non-lethal energy weapons are also finding applications on

a larger scale, in terms of what's known as "area denial." This is used in crowd control. Say there's an area—such as a VIP's home—that police want to keep an angry crowd out of at all cost. One option (albeit rather unpleasant) is to ring-fence the area with microwave generators. "Imagine how you would feel if you were inside a microwave oven," says Michael Wartell. "You would feel as if your skin was cooking. So if you can direct microwaves at specific areas you can cause anyone in those areas a great deal of pain, so that they don't want to be there."

Experimental area-denial systems using microwaves are now being developed by military organizations all over the world. Wartell says that in addition to crowd control, they could be used for flushing out hidden snipers.

Another energy technology being looked at for area denial is acoustics—sound beams. XADS has developed its Screech "acoustic disruptor," capable of giving anyone within 20 meters a severe headache in seconds.

In 2004, it was reported by the *Washington Times* that U.S. troops in Iraq were using similar sonic weapons known as LRADs (Long Range Acoustic Devices). The weapons were developed in response to the terrorist attack on the USS *Cole* and designed to be mounted on ships to give protection against attackers in small boats. The weapon itself measures almost a meter across and emits a high-pitched, piercing sound beam at a volume of up to 150 decibels (a smoke detector alarm is typically 80 to 90 decibels). The sound causes extreme pain and has an effective range of between 100 and 600 meters. A report published by Wired News in September 2005 suggested that the U.S. Navy had 60 of the devices in service at that time, and that American police and law enforcement organizations were also beginning to show an interest.

In *Doctor Who,* sonic weapons are a popular choice with many alien races. In the Second Doctor serial "The Ice Warriors," the eponymous Martians use guns that they claim will "burst your brain with noise." (Incidentally, the Ice Warriors also had larger versions of these sonic weapons fitted to their spacecraft. You might wonder quite what use sound weapons are to them in the vacuum of space, where it's commonly thought that sound cannot travel, but "in fact, sound can travel in space—it's not a perfect vacuum," says Gerry Gilmore of the University of Cambridge's Institute of Astronomy.)

More recently, in the Ninth Doctor series, the Doctor's colleague Captain Jack Harkness is a self-confessed fan of sonic blasters, cannons, and disruptors. These are depicted as highly lethal shooters—no messing about with non-lethal area denial here. Some researchers think that high-power sound does indeed have the potential to create killing weapons. "At above 150 decibels sound is said to be deadly," says James Friend, an ultrasonics researcher at Monash University in Australia. "I have worked with ultrasonic transducers that generated in excess of 180 decibels, but at a high enough frequency to avoid danger."

Friend thinks that such a weapon—in the present day, at least—would need to be much larger than a rifle or pistol, probably with a large ungainly shaped component at the front in order to transmit the sound into the air. This is the tricky part with building sonic weapons. The tendency is for sound waves to remain trapped in the denser material of the gun and not enter the thinner medium of the air. However, Friend cites the recent deployment of sonic weapons in Iraq as a clear sign that progress is being made in this area.

Perhaps the loudest sonic weapons are nuclear explosions, although it's very hard to say exactly how loud they are. This is because any sound louder than about 200 decibels ceases to behave like a sound wave and acts more like a supersonic shock wave instead. Estimating the loudness of a shock wave is extremely complicated. And, at any rate, if you were that close to a nuclear explosion, losing your hearing would probably be the least of your worries.

The Daleks and the Cybermen are especially fond of nuclear weapons. But they're not the only ones. In the Fourth Doctor adventure “Underworld,” the Doctor meets a race armed with “fission grenades”—handheld nuclear devices with a yield of 2,000 megatons.

Are handheld nuclear weapons feasible? In reality, perhaps the smallest nuke ever deployed in the field was the US M-388 Davy Crockett nuclear bazooka. It was designed as a tactical artillery piece for holding back advancing Soviet troops in Europe and consisted of a recoilless bazooka tube about 15 centimeters in diameter, mounted on a tripod. The size of the warhead was very much less than 2,000 megatons, though—it had a variable yield that could be set anywhere between 10 and 250 tons. But even with this small yield you’d have to be mad, or very dedicated, to use one of these things. That’s because the Davy Crockett had a range of just 4 kilometers (2.5 miles)—less than the blast radius of the resulting nuclear explosion.

The Davy Crockett was certainly nowhere near handheld. The complete projectile was nearly 80 centimeters long and 30 centimeters across, and weighed 35 kilograms. The weapons were carried by the US armed forces between 1961 and 1971.

The Davy Crockett was a fission device, releasing energy by splitting the nuclei of heavy atoms. According to nuclear weapons expert Frank Barnaby, a member of global security organization the Oxford Research Group and author of the book *How to Build a Nuclear Bomb,* it’s impossible to build a fission weapon small enough to fit in your hand. Indeed, any fission bomb much smaller than the Davy Crockett would be unable to hold the necessary “critical mass”—the minimum amount of uranium or plutonium needed to trigger a nuclear explosion.

But there’s another way to create a nuclear explosion—fusion, whereby energy is released by joining together lighter atomic nuclei. Doing this requires heating the nuclei to terrific temperatures—around 100 million degrees centigrade. Barnaby says that a small fusion device could be made by surrounding a medicine-bottle-sized flask of tritium—a type of hydrogen in which each atom has two neutrons added to its nucleus—and then compressing it to enormous temperatures using a high-explosive charge. It would be feasible to make a fusion device with a yield of a few tens of kilotons this way. The device would measure about the size of a football—just about handheld, assuming you're allowed to use both hands.

The biggest nuclear device ever detonated on Earth was a 57-megaton fusion bomb, known as the Tsar Bomba, exploded by the Russians in a test on October 30, 1961. Could this enormous yield ever be upped to the 2,000 megatons of the fission grenades from *Doctor Who*? "It would be laughed out of court," says Barnaby. "Theoretically you can have bigger bombs, but if they're not militarily deliverable—if you can't get them in a plane or on a missile, then what's the point?"

Operational warheads today are already quite huge. The submarine-launched Trident missile has a warhead about 3 meters long. The Tsar Bomba was 8 meters long by 2 meters wide—so big that the bomb-bay doors had to be removed from the aircraft that delivered it.

In Fourth Doctor serial "Underworld," two fission grenades are used to destroy a planet. However, as we saw in Chapter 7, a 2,000-megaton nuclear bomb would be grossly underpowered for this task. Even with two such bombs, they would fall short in destructive power by a factor of about 10^{13}, that's 10,000 billion.

Weather Wars

One of the more bizarre weapons used to threaten the Earth by an alien race was its own weather. In the Second Doctor adventure "The Moonbase," the Cybermen attempted to take control of an outpost on the Moon, used for controlling weather on Earth in the year 2070. Their plan was to manipulate the weather and use it to destroy all life on the planet.

Humans used the device, called the Gravitron, for controlling potentially dangerous weather—tempering heat waves and steering hurricanes safely out of harm's way. In fact, this isn't so daft. In the wake of Hurricane Katrina, which destroyed much of New Orleans in late August 2005, scientists have suggested that hurricanes might really be controlled, or even dissipated, by bombarding them with microwaves from space.

Ross Hoffman, working for NASA's Institute for Advanced Concepts, has carried out a number of computer simulations showing how small changes in the conditions prevailing as a hurricane forms can drastically alter the way the final storm turns out. In particular, he found that temperature changes of just 2 to 3°C can have a major effect on a hurricane's course.

Hoffman envisages solar-powered satellites that could fire microwave beams down into incipient hurricanes from Earth orbit. Because the hurricanes are made up of clouds of water vapor, then just like a bowl of soup in your microwave oven, they'll readily absorb the microwaves and convert the energy into heat, warming the clouds up. The technology to do this is still in its infancy, but Hoffman believes it's eminently possible.

Could the Cybermen turn such a device into a weapon? Quite possibly—however, they would be in breach of international—or even intergalactic—law. As Hoffman pointed out in an interview for CBS News: "There is a UN convention in place that bans weather modification as a weapon."

Not content with stirring up its weather, some especially belligerent aliens have decided to have a go at knocking a planet's parent star for a loop instead. In the Third Doctor serial "Colony in Space," we learn how the super-advanced ancient inhabitants of the planet Uxarieus built a doomsday weapon capable of making a star explode—and that the Crab Nebula, 7,000 light-years from Earth in the constellation of Taurus, is the remains of a star that they tested the weapon upon.

Back in the real world, astronomers today know that the Crab Nebula is indeed the remains of a star that exploded—it's the remnant cloud from a supernova explosion that was observed and recorded by the Chinese in the year 1054. A supernova is what happens when a large star reaches the end of its life (see Chapter 31). When that happens, it's curtains in a big way for any planets orbiting the star. But could an advanced civilization ever have the means to inflict one artificially?

"I cannot see a physical reason why this could not happen," says astrophysicist Martin Barstow at the University of Leicester. "Although practically it would be rather hard."

Barstow points out something that astronomers have known for years: big stars burn fast. A star with the mass of the Sun has a lifetime of about 10 billion years, whereas one that's ten times more massive will burn through its fuel 1,000 times more quickly and so live only for about 10 million years before going supernova. So if you can pile enough mass onto a star, you can in principle make it go supernova much sooner than it would normally.

Gerry Gilmore thinks that this is way beyond human capabilities for the foreseeable future. "Our Sun annihilates 600,000 tons of matter every second to generate its heat and light," he says. "One would have to put substantially more than this mass into the Sun to have even a tiny effect. Very much larger effects are needed to generate a supernova." Indeed, in order to bring a star's lifetime from billions of years down to a few million you would need to add about ten solar masses' worth of material.

A star's energy is generated deep inside, in its core. And so another way to crank up the burn rate might be to fiddle with its core in some way—perhaps squeezing it or heating it up. One option might be to fire neutrinos into it. These are ghostly subatomic particles that are produced in abundance by the nuclear reactions taking place in the Sun's interior (we know this because detectors on Earth pick up a huge amount of them streaming out of the Sun). Neutrinos pass through matter very easily, meaning that they could be beamed through the Sun's outer layers with ease but would then be intercepted by its dense core region. Neutrinos play a key role in the physics of the solar core and so dumping a large quantity of them in there could conceivably speed up nuclear reaction rates, making the star reach the end of its life and explode sooner.

But Andrew King, an astrophysicist at the University of Leicester, thinks that even this plan is doomed to failure, for two reasons. Firstly, the timescales involved are immense. King has calculated that it would take around 30 million years for an effect introduced in the core to have a marked influence on the rest of the star. And secondly, stars are such stable objects that destabilizing them requires huge amounts of energy—so much that you may as well just fire it all at the planet you're attacking in the first place, rather than waste your time mucking around with its parent star.

"Stars are extremely stable objects," says King. "If it were possible to destabilize them easily, as these wonder weapons seek to, it would have happened naturally long ago, thanks to the large neutrino flux being given off by nearby stars. Then there would be no stars, no life and no *Doctor Who*. So it doesn't look promising!"

Not for the aliens, at least. But for us here on the potential receiving end, that's all rather jolly. Now all we need is something to help fend off all those alien laser cannons, sonic disruptors, and nuclear hand grenades that we met earlier on. What we want are some deflector shields.

28

Force Fields

"Come on out. It's okay—that force field can hold back anything."
"Almost anything."
"Thanks, I wasn't going to tell them that."
"Sorry."

—Captain Jack Harkness apologizes to the Ninth Doctor after Dalek death rays bounce harmlessly off the Tardis's force field in "The Parting of the Ways"

Force fields are just one of a number of alien technologies featured in *Doctor Who* that enable the user to influence objects from a distance—be they incoming projectiles, laser blasts, or hapless assistants dangling from dirigibles. And they're not quite as fanciful as you might imagine.

In Chapter 10, we saw how military engineers are already building something similar to the bullet-dissolving force fields employed by the Daleks in the 2005 series. But this "electric armor" can work only when its outer metal layer is earthed—otherwise the electrical current that it relies on has nowhere to flow to. If you're out in space, light-years from any suitable pipework to grab hold of, then that's going to be a problem. So how might the deflector shields around spacecraft such as the Tardis actually work?

Feel the Force

Some researchers have mused on the idea of using gravity to bat away hostile projectiles and laser blasts. Einstein's theory of general relativity (see Chapter 3) says that gravity is just curvature of space and time. So by setting up a strong enough gravitational field around a spacecraft, you might think it's possible to curve away the trajectories of incoming objects.

One problem: gravity is an attractive force. It pulls apples down to the ground—it doesn't make them fly off into space. So if you're not careful, raising deflector shields of this sort could get you shot even more often.

But what if it were possible to generate some kind of repulsive gravitational force—antigravity? The gravity that we're used to is attractive because it's generated by objects that have positive mass. If you wanted to generate some antigravity, you'd need material with negative mass. As we saw in Chapter 2, scientists have already succeeded in doing this in the lab—in a phenomenon called the *Casimir effect.* Could we then use the Casimir effect to generate a large enough negative mass to power our spacecraft deflector shield?

Physicist Jim Al-Khalili, of the University of Surrey in England, isn't convinced. In fact, he believes that any attempt to build a gravitational deflector shield is fundamentally flawed. The trouble is that gravity is the weakest of all the forces of nature. It takes all the gravity produced by the entire mass of the Earth just to stick our relatively tiny and slow-moving bodies to the planet's surface. If you want to deflect fast-moving missiles and laser beams, you're going to need a device capable of generating a much brisker force than that—and you're probably going to want it in a box several orders of magnitude smaller.

So gravity's out. There are three other forces of nature that we can try: electromagnetism and the strong and weak nuclear forces. Of these, the strong nuclear force is the most powerful. It welds together protons and neutrons in the nuclei of atoms with such force that breaking them apart releases stupendous quantities of energy (it's the force responsible for nuclear power). But this force is also extremely short ranged, falling to zero in a distance less than the diameter of an atom. That means that deflector shields based on the strong nuclear force—like maybe the "neutron force field" used to thwart the Cybermen in "The Wheel in Space"—are something of a non-starter.

Al-Khalili puts his money instead on electromagnetism—the force responsible for creating electric and magnetic fields and the one that also binds together molecules. It's much stronger than gravity, so that substantial fields can be generated from a relatively compact device. And it has a much longer range than the nuclear forces. The one problem is that electromagnetism can influence only bodies that are electrically charged. But Al-Khalili thinks he has the solution to that.

He proposes bombarding the target with a beam of positrons to charge it up. Positrons are the antimatter equivalent of the electron and are given off by certain atomic nuclei in a process known as *positive beta decay.* Whereas electrons have negative charge, positrons are positively charged. And when the two types of particle come together they totally annihilate one another. Al-Khalili thinks that these properties can be used to tilt the balance of charge in an object that you want to deflect. "You can use positrons to destroy electrons in the target," he says. "And if you destroy enough of them, then the target becomes positively charged. Then you can whack on an electric or magnetic field to deflect it."

The idea could also explain tractor beams and their opposite number, repulsor beams. These are devices that can pull or push objects from a distance. A tractor beam was used to haul Mavic Chen's spacecraft back to the planet Kembel in the First Doctor story "The Daleks' Master Plan." And when Rose Tyler falls from a blimp in the Ninth Doctor episode "The Empty Child," Captain Jack Harkness catches her with a repulsor beam from his ship. Jack even tells Rose to switch off her cell phone as it's interfering with the beam—suggesting that the device really is electromagnetic in nature.

Blocking Beams

That's all well and good when you're dealing with solid objects and projectiles, but how do you go about fending off the blast from a beam weapon such as a laser? A laser beam is made up of the same kind of particles—photons—that comprise the electromagnetic field. So you might think that deflecting one with the other would be a piece of cake. Yet photons don't readily want to interact with one another.

There is a way out, although it could be a long shot. It comes in the form of an effect known as *Delbrück scattering*. Quantum theory says that subatomic particles, photons of light included, can very briefly turn into other particles, so long as certain rules are obeyed—such as that the energy used to create the new particles is the same as the total energy of the ones that went in, that the electric charge is the same, and so on. In Delbrück scattering, a photon of light turns briefly into an electron-positron pair. These charged particles can then be deflected by an electric field before they recombine back into a photon—now traveling in a slightly different direction.

Predicted by German physicist Max Delbrück in 1933, the effect was confirmed experimentally by Robert Wilson, working at Cornell University in 1953, when he measured the deflection of light by the electric field surrounding the nuclei of lead atoms. Classical (that is, non-quantum) physics already predicts that light will be deflected as it bounces off an atom—an effect known as *Thomson scattering*, after the English physicist J. J. Thomson who first explained it. What Wilson found was a tiny deviation of just a few percent from Thomson's effect that was in good agreement with Delbrück's calculations. The effect is small but it means that electric fields really can deflect light beams.

Remember that next time a Dalek points a ray gun at you.

29

The Matrix

"A man is the sum of his memories . . . A Time Lord even more so."

—The Fifth Doctor, "The Five Doctors"

In 1976, a *Doctor Who* serial called "The Deadly Assassin" introduced viewers to a rather interesting concept: the "Matrix." This was back while Keanu Reeves was still knee-high to a flash card, but the idea was very similar to that put forward in the cyberpunk movie series in which he would star 23 years later.

The Matrix of *Doctor Who* is a virtual reality world, based inside a massive computer system on the Time Lord home planet of Gallifrey. Time Lords access it using either a special headset that links the computer to their brains or by walking through the Seventh Door, which the Doctor discovers in the 1986 adventure "The Ultimate Foe."

In "The Deadly Assassin," the Doctor enters the Matrix to fight Gallifrey's Chancellor Goth (who has been corrupted by the Doctor's archenemy, the Master). Goth has mastery over the virtual world, allowing him to manipulate it and to bend or break the laws of physics within it.

The Matrix was also home to a sub-system known as the Amplified Panoptric Computations Network (or APC Net). As well as storing each Time Lord's unique "biological imprint," the APC Net regularly received data from every Tardis about each Time Lord's activities and acted as a repository for all of the knowledge ever amassed by the people of Gallifrey. The idea was that at the moment of a Time Lord's death, an electrical scan would be made of his brain so that all of his memories and experiences would be stored for posterity.

The idea sounds very similar to a collaboration between neuroscientists, psychologists, and computer experts in the UK known as Memories for Life. "Being able to store a person's memories electronically is exactly what we're talking about," says Wendy Hall, Professor of Computer Science at the University of Southampton and a member of Memories for Life. "Imagine having a video of your entire life. It might be rather boring, but you could do it. It's not a problem to store 70 years' worth of video data using today's technology."

How would you do it? Linking computer hardware to our brain's "wetware" is fraught with difficulty. There are roughly 10 billion neurons (nerve cells) involved in the storage and manipulation of information in the human brain, each of which develops around 1,000 connections to neighboring cells, the precise configuration of which is how the brain stores its memories. Quite how a computer would get at that information without resorting to lethal invasive techniques isn't yet clear. And the best we can do non-invasively at the moment is to use an fMRI scanner, the resolution of which is poor, picking up details no smaller than about a millimeter in size.

But Hall doesn't think this will matter. She argues that if we want to gather and store our experiences, then it's far easier to do this using devices external to the brain, rather than to let the brain do the recording and then have to try to suck the information out somehow. The technology is already here for cameras, microphones, and miniaturized computer hard drives to unobtrusively record our memories in tandem with the brain. For example, in 2005, Fujitsu released an 80-gigabyte hard drive that's just 2 centimeters across.

Even though these devices won't link into our heads directly, Hall believes that we will quickly come to think of them as natural extensions of our brains. "You have glasses, you have hearing aids and in the future we'll have memory aids," says Hall. She isn't yet sure exactly what the devices will look like. But given that sound and vision will be key elements, it seems reasonable to suppose they might take the form of a pair of spectacles incorporating a camera and a microphone to see and hear everything we do. Images could then be displayed on a tiny screen on the lens and sounds played back via a small earpiece.

"So when you go up to someone and you can't remember their name you'll have a little memory aid, speaking into your ear, telling you this is Mr. So-And-So who you met last week," she says. Already such technology is becoming available. In October 2009, Microsoft announced the SenseCam—a camera worn around the neck which can be set to take pictures at regular intervals to document the user's life.

Future applications could be even more powerful. For example, do you ever find yourself racking your brain, trying to think where it was that you saw or heard a certain piece of information? Let's say it was the details of a discount vacation. Where was it? A newspaper article? A TV ad? Did a friend tell you about it? Imagine, then, if you could ask your memory aid to search every experience you've had in the past week—everything you've read, everything you've watched on TV, every conversation you've had—looking for the word "vacation."

Memories for Life researcher Kieron O'Hara, a computer scientist at the University of Southampton in England, likens this to the emergence of writing. As we moved from being an oral to a literate society, we didn't need to remember so much. Effectively we used pen and paper to outsource our thoughts. Writing, he says, was largely developed for us to use as proxy memory, leaving our biological memory free for remembering other things. Now, he says, Memories for Life technologies, and already the Internet, are doing the same thing but in a much more powerful way. "Why should I bother to remember a boring work conversation when I could record it, have it automatically annotated, and render it searchable with clever language-recognition technologies?" he says. "If I wanted to hear the conversation again, I would simply search for it in my 'memory' and replay it, exactly in the same way as I can reread old letters or emails."

One issue that Memories for Life is looking at is ensuring that data written today will be accessible hundreds or maybe even thousands of years into the future. I'm sure I'm not alone in having a whole rack of diskettes sitting on my shelf, which, since Apple stopped fitting floppy disk drives in their Macintosh computers, I now have no way of reading—nor probably ever will. So there's the format problem. There's also the problem of the physical durability of some media. Even data stored on a CD-ROM typically won't last for longer than 5 to 10 years, and magnetic storage devices, such as tapes and hard drives, suffer similarly. If Memories for Life is to work, media will need to be extremely durable and long-lived.

"This is the $64,000 question," says O'Hara. "Something like the current identity card scheme might well have to keep tabs on people who will live until, say, the year 2100—in other words, twice the entire history of computing!" Making electronic storage media that can last this long isn't easy.

But, nevertheless, a lot of companies are giving it a go. A firm called Norsam Technologies based in Santa Fe, New Mexico, has developed what it calls its HD Rosetta archival preservation system, which works by etching data onto extremely durable nickel-based disks. The company claims that these are immune to technology obsolescence, water damage, and electromagnetic radiation, as well as being highly resistant to temperature variations. The British Library is also looking into long-term preservation technologies for its digital records.

Alternatively, O'Hara suggests that the World Wide Web could become an information repository, where data could be parked and accessed from anywhere. Digital information could then take on an almost organic existence on the Web, outlasting any particular storage medium as it's copied from computer to computer in much the same way that genetic information outlasts its biological hosts as it's passed on from one generation to the next.

And what about the format problem? "For long-term readability in such a world you would expect to lose many of the clever coding devices that need special software to be read," says O'Hara. "Instead you'd have clear, open standards about how to store info that are very adaptable. So even if your info was in an out-of-date format, it wouldn't be beyond the wit of man to write a translator program from the older format to a newer one." So maybe there's hope for my old diskettes yet.

Another issue is privacy. Civil liberties are being increasingly curtailed in the name of law enforcement and national security. If you were implicated in a serious crime, one of the first things the police would do is to seize your computer to search its hard drive for evidence. Would this also be the case with your memories? Would there be any limits on the memories that authorities would have the right to access?

One big problem for any government agency wanting to do this is the huge amount of information they would have to trawl through. Normally when faced with such a vast amount of data, a researcher would skip straight to the index. But O'Hara thinks that this will be the issue that really makes life hard for the spooks: the data will be filed using an index that only the owner of each memory will really understand.

He likens the situation to Marcel Proust's novel *Remembrance of Things Past,* in which the taste of a Madeleine cake enables the narrator to experience his childhood memories in the French town of Combray. "If his memories were indexed to be associated with the taste of a Madeleine (and not just shoved in a folder called "Childhood Days in Combray"), then they would be all the more difficult for outsiders to get hold of as they would not have the correct access to the right associations," says O'Hara.

That means that any law enforcement agent trying to find out who you were with at 10 p.m. last Friday might have quite a job skipping straight to the information they're after—and may be faced with the gargantuan task of having to sift through every single memory you have stored.

The Matrix is used for this purpose in the Sixth Doctor "Trial of a Time Lord" series of linked adventures. The Doctor is tried by his fellow Time Lords on Gallifrey, charged with breaking the first law of time—which says that a Time Lord must never meet his former or future selves (see Chapter 3). Much of the evidence presented through the Matrix, however, has been tampered with by the Doctor's prosecutor, the evil Valeyard.

Indeed, if stored memories do ever become admissible as evidence in court, then the Doctor's exploits serve as a cautionary reminder that electronic evidence isn't always as reliable as it's cracked up to be.

• Part Four •

MISSION TO THE UNKNOWN

30

Event One

"He got it wrong on the first line! Why didn't he ask someone who saw it happen?"

—The Fourth Doctor reading the book *Origins of the Universe* by Oolon Caluphid, in "Destiny of the Daleks"

To answer the Doctor's question above: probably because anyone who was there would have been burnt to a crisp and squashed flat. If the Doctor, or anyone else, had been present shortly after the Big Bang, in which our Universe is thought to have been born, he or she would have had to endure temperatures on the order of 10^{32} °C. That's a number 1 with 32 zeroes after it, 10^{30} times hotter than boiling water: a bit warm. The density of matter at that time was believed to be an even more staggering 5×10^{96} kilogram per cubic meter—while the pressure was off the scale at around 10^{110} times the atmospheric pressure at the surface of the Earth. All in all, it's probably more than the Tardis's deflector shields (see Chapter 28) could have coped with.

And even if the Doctor had been there, he wouldn't have seen very much. Up to about 300,000 years after the Big Bang, the Universe was completely opaque to electromagnetic radiation—of which the light that we use to see is one type. That's because up until this time the matter in the Universe was ionized. All of its atoms were stripped of their electrons, forming a sea of positively and negatively charged particles which scattered photons of radiation in all directions.

Nevertheless, thanks to a combination of astronomical observations of the present-day Universe and ingenious theoretical inference, scientists have managed to piece together an amazingly precise picture of how our cosmos was probably born and grew up.

In the Beginning

The story begins about 13.7 billion years ago when a bubble of space and time spontaneously popped into existence and started to expand. Before that time there was nothing, no matter, no space—not even time. Indeed, asking what happened before the Big Bang has famously been likened to asking what lies north of the North Pole.

It was physicist Albert Einstein who first realized that the Universe could appear from nothing in this way—and he almost got himself killed in the process. It was one day during the 1940s and Einstein was out walking with his Princeton colleague George Gamow, when Gamow mentioned that one of his students had calculated how it's possible to make a star from nothing because its mass energy is exactly equal but opposite to its gravitational energy. Einstein, realizing immediately that the same principle could apply to the Universe at large, stopped in his tracks—in the middle of a busy road, forcing several cars to swerve in order to avoid hitting them.

Creation from nothing is a phenomenon that's commonly encountered in quantum theory, thanks to Heisenberg's uncertainty principle, which we came across in Chapter 2. Formulated by the German physicist in 1927, it says that you can never know the speed and position of a subatomic particle—such as a proton or an electron—at the same time. No one really knows why this should be, but it certainly appears to be true and the principle is now a cornerstone of quantum physics. Heisenberg went on to show that linking position and speed in this way is equivalent to linking mass and time—in the quantum world, subatomic particles can pop in and out of existence so long as their masses and the times that they exist for remain within the bounds of the uncertainty principle. The effect is confirmed by experiments.

During the 1970s, physicist Edward Tryon of the City University of New York suggested that, just as quantum uncertainty brings virtual particles into existence, it could also have given birth to the tiny seed from which our Universe then grew—providing the mechanism to realize Einstein's insight.

But there was still a problem. Packing the Universe into something the size of a quantum particle would have created a gravitational field so strong that it should all have recollapsed, crushing the Universe out of existence again in an instant. Something must have happened to the embryonic cosmos to blow it up out of the quantum realm before it had a chance to fall back in on itself. But what?

In the late 1970s and 80s, cosmologists led by Alan Guth at MIT put forward a theory which said that roughly 10^{-35} of a second (the number 1 divided by a 1 with 35 zeroes after it) after the Big Bang, the Universe entered a phase of super-rapid expansion called "inflation." The Universe inflated until it was 10^{-32} of a second old, and during this time increased in size by a factor of 10^{30}, emerging at the end of the inflationary epoch roughly the size of a grapefruit. Inflation puffed the Universe up sufficiently that there was no chance of it recollapsing. (Although I say above that inflation happened after the Big Bang, the timing depends on what we really mean by the term "Big Bang." If we define the Big Bang as the moment that our Universe was created, then the above is correct. However, some cosmologists argue that we should really reserve the name Big Bang for the fireball created at the end of inflation—see below. And in that case, inflation actually took place before the Big Bang.)

Inflation is thought to have been caused by certain kinds

of subatomic particles in the early Universe, known as *scalar fields*. These are defined as particles that lack a quantum property known as spin. Most subatomic particles have spin. It's a number assigned to each type of particle that helps to determine its behavior according to the mathematics underpinning quantum physics. It's broadly analogous to what we think of as spin in the everyday sense, but with a few unusual extra features. For example, in the early decades of the twentieth century, physicists discovered that particles with spin values of one half, three halves, five halves, and so on, obeyed a different set of quantum laws to those with whole-number spins, such as 1, 2, 3. Scalar field particles, however, have zero spin, making them the simplest kinds of particle that can exist.

During its early history, as we saw in Chapter 11, the Universe is believed to have gone through a number of "phase transitions"—wholesale shifts in its state, rather like the phase transitions that occur when steam cools and turns into water. During a cosmic phase transition, a scalar field drops from a high-energy state to one of lower energy. But this didn't occur at exactly the same time everywhere in the Universe. In some places it would have happened slightly earlier; in others slightly later. So in some patches of the early Universe, small pockets of scalar field particles would have remained trapped in a high-energy phase, while all the particles around had slipped down to low energy. In this case the pocket, and the very space that contained it, would have inflated rapidly—rather like an expanding bubble in a pan of boiling water. After about 10^{-32} of a second, the pocket would have fallen into the lower energy phase, at which point inflation ended and the scalar field decayed into a hot soup of matter and radiation that went on to form the Universe that we see today. Figure 14 offers a description of how the Universe is thought to have expanded: exponentionally during inflation compared with the linear expansion it's undergoing today.

Scientists think that inflation could explain a number of the shortcomings in the traditional Big Bang model. For example, why does the curvature of the Universe seem to be so flat? Infla-

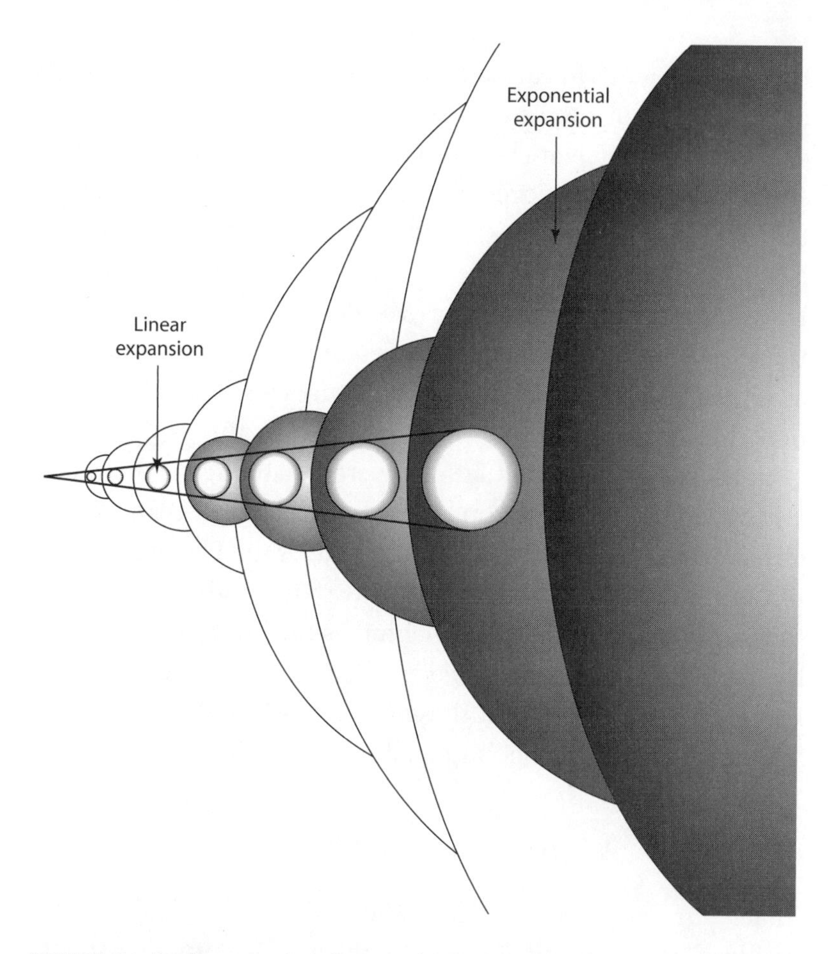

FIGURE 14. During cosmic inflation, the early Universe is believed to have expanded exponentially. Today, the Universe is expanding linearly—that is, it's expanding by the same amount each year. During exponential expansion, however, the size of the Universe increases by a constant multiplicative factor every year—so it might double, or even expand by a factor of ten each time. This figure shows how exponential expansion (*outer spheres*) can soon lead to a much bigger universe than linear expansion would (*inner spheres*).

tion solves this neatly, making the Universe so darn big that it just appears flat—in the same way that the surface of the Earth appears flat on small scales. Another problem was: why do opposite sides of the observable Universe look the same, even though they've never actually been in causal contact with one another? Inflation neatly dodges this one by saying that these regions of the sky were once in causal contact, but were then blown apart dramatically during the inflationary phase transition.

It could even explain how the galaxies were born. These massive collections of hundreds of billions of stars are now thought to have condensed gravitationally around small density irregularities in the early Universe that were left behind by inflation. It works like this. As inflation proceeded, the scalar field driving the expansion would have undergone quantum fluctuations, as all subatomic particles do in accordance with Heisenberg's uncertainty principle. However, because of the rapid expansion of space, these fluctuations were blasted from the quantum world up to vast astrophysical length scales. Astronomical observations of the large-scale structure of the Universe and of the lumps and bumps in the microwave background radiation—the electromagnetic echo left behind as the intense radiation of the Big Bang fireball expanded and cooled—show a good agreement with the theory that these features of the Universe were seeded by quantum fluctuations during inflation.

The End of the Universe

That's our best theory for how the Universe and the galaxies within it were created. So how might it all eventually come to an end, if indeed it does? This is a little more contentious. It all depends upon the density of the matter filling the Universe today. If it's above a certain threshold, then there's enough gravity to make the Universe ultimately recollapse back down on itself. If this happened, then at some point billions of years in the future, astronomers would notice the expansion of space begin to slow down—as the entire history of the cosmos started to play itself back in reverse. Galaxies would stop rushing apart and start converging back toward each other. The microwave background radiation that has been spreading out and cooling down since the moment of the Big Bang would begin to reheat as the Universe collapsed. Eventually it would become hotter than the stars, causing them to break up and evaporate. Galaxies would collide and overlap, and matter would be devoured by the giant black holes lurking in their centers. Finally, the Universe would squeeze back down to subatomic size and, just as quickly as it was born, it would be gone.

Cosmologists call this scenario the Big Crunch. Some believe that if it happens there may be a final optimistic twist to it. Rather than snuffing itself out of existence, the Universe could bounce back, rising phoenix-like from the ashes to begin a new phase of cosmic expansion.

Whether the Universe will really end in a Big Crunch depends on the value of a number that cosmologists call the cosmic density parameter—Omega, Ω—defined as the actual density of the Universe divided by the "critical" density needed to make it recollapse. So if the Universe has exactly the critical density, then $\Omega = 1$. The trouble is that the value of Ω depends not just on the quantity of matter in the Universe but also on the rate at which space is expanding. Recent results from NASA's Wilkinson Microwave Anisotropy Probe (WMAP) give $\Omega = 1.02$ plus or minus 0.02—tantalizingly close to 1, but too close to call with any degree of certainty.

The Big Crunch corresponds to the case in which Ω is bigger than 1. But what if Ω is less than or equal to 1? The good news in that case is that the Universe won't recollapse. But little good will it do us. Instead, space will just expand for eternity, the density of the matter and radiation within it becoming gradually diluted away to zero. The stars will burn out and die. The galaxies will slowly switch off. And eventually even fundamental particles of matter will decay away, leaving just empty space in all of its vast inky blackness. This scenario for the fate of the Universe is known among cosmologists as the Heat Death.

Some physicists like to view the Heat Death in terms of a quantity they call *entropy*. This is a measure of the degree of disorder in a physical system. For example, a tidy desk with books and papers all neatly filed and stationery all put away in drawers and holders has a low entropy—everything is in an organized state. On the other hand, a messy desk with items strewn randomly across it has high entropy. The Heat Death can be viewed as the Universe reaching its state of maximum entropy. Rather like an untidy desk with detritus spread evenly across it, everywhere in the maximum-entropy Universe has the same temperature—hot spots, such as stars, and cool regions, such as empty space, are all smoothed out into a uniform lukewarm background, gradually getting cooler as the cosmos keeps on expanding.

Entropy is a recurrent theme throughout the Fourth Doctor's regeneration adventure "Logopolis." In this story, the Heat Death of the Universe has already started to happen, and it's only the continuous calculations being carried out by the inhabitants of the planet Logopolis that are keeping it at bay. They're calculating how to hold open wormhole-like structures in space (see Chapter 3), down which the entropy of the Universe is being drained away. Meanwhile, the Doctor's enemy, the Master, is trying to disrupt their calculations and so bring about the Universe's destruction.

John D. Barrow, a cosmologist at the University of Cambridge, is doubtful whether the Logopolitans' ploy could really work. "The amount of work that has to be done identifying and funneling the high-entropy material down the wormholes may well create more entropy than it saves," he says.

Barrow believes that a better method to avoid the Heat Death is to make the Universe expand faster in some directions than in others. Having different expansion rates in different directions sets up temperature gradients (effectively restoring some sense of hot spots and cold spots) and so moves the Universe out of its state of maximum entropy. Quite how the Logopolitans would go about engineering such skewed expansion across billions of light-years of space is not at all clear. Although if they're capable of making entropy-draining wormholes into other universes and other dimensions, then perhaps this will be a cinch for them after all.

More recently, scientists have come up with a third possibility for the ultimate fate of the cosmos. It stems from the discovery by astronomers during the late 1990s that the Universe isn't just expanding, but that the expansion seems to be accelerating. A team led by Saul Perlmutter at Lawrence Berkeley National Laboratory in California found observational evidence that space is permeated by a kind of antigravitational material known as "dark energy," that's making the Universe expand ever faster. It's somewhat reminiscent of the exotic matter that we encountered in Chapter 2. In 2003, a group of theoretical cosmologists led by Robert Caldwell of Dartmouth College, New Hampshire, took Perlmutter's discovery further, calculating that if the form of the dark energy was potent enough, then it could eventually accelerate the expansion of the Universe so violently that it would literally tear everything in it apart.

Caldwell named his new cosmic catastrophe the Big Rip, while the particular form of dark energy needed to bring it about was called *phantom energy*. The best astronomical observations—such as those made using the WMAP probe—aren't yet accurate enough to say for sure whether or not phantom energy rules the Universe, but they certainly don't rule it out. The good news is that if our Universe is destined to end this way, it won't happen for another 20 billion years.

It's just as well we have a little time. When the Big Rip strikes, it will be a force to be reckoned with—and even the Logopolitans may have their work cut out stopping it.

31

The Eye of Harmony and Other Black Holes

"I wonder what it would be like to be crushed into a singularity."
"Don't stand there wondering; do something!"

—The Fourth Doctor and Romana, "The Horns of Nimon"

A black hole is a region of space where matter has become so dense that its gravitational field is too strong for anything, even light, to escape from it. Anything passing over the outer boundary of a black hole can never escape and is doomed to be crushed out of existence in the hole's core, a point-like "singularity" where the laws of physics break down.

As we'll see, the Doctor has encountered black holes on several occasions—not a surprise, as they're thought to be remarkably common in the Universe.

Perhaps the most numerous black holes are those formed by the deaths of massive stars in supernova explosions. These are colossal explosions, each marking the death of a high-mass star. During a supernova, the star temporarily flares so brightly that it outshines the combined light from all the 100 billion other stars in its home galaxy. The crushing force of the explosion compresses the star's core. For lighter stars, less than about 3.5 times the mass of the Sun, this force packs negatively charged electrons and positively charged protons together to form a big ball of zero-charge neutrons—a neutron star, weighing a few times the mass of the Sun but measuring just 25 kilometers across.

Neutron stars are supported by what's called *degenerate neutron pressure.* In a normal star, like our own Sun, the tendency of gravity to pull the star in on itself is balanced by thermal pressure pushing outward. In a neutron star, however, gravity is much stronger and so a much stronger force is needed to balance its inward pull. Degenerate neutron pressure does the job. This is a quantum mechanical force, arising from the fact that neutrons tend to resist being squeezed into the same quantum state as one another.

But this works only if the star going supernova is lighter than about 3.5 solar masses, otherwise the gravity of the core is too great for even degenerate neutron pressure to support it. No known force of nature can halt the gravitational collapse (a possible exception are quark stars—see Chapter 33). In this case, theory predicts that the star's core shrinks down to become a point of zero size and infinite density, known as a singularity, where the strength of gravity becomes infinite and the laws of physics cease to apply.

The singularity is surrounded by a spherical surface called the *event horizon.* No light can escape from within this boundary and so it appears black—this is where the name "black hole" comes from. If the Sun could be packed down to form a black hole, its event horizon would be about 6 kilometers across.

There's good evidence for the existence of these dead-star black holes. Although black holes can't be seen directly, occasionally they exist in binary systems where their gravitational effect shows up through its influence on their bright companion stars.

Physicists sometimes draw the analogy with ballroom dancers—the men are difficult to see in their dark suits, but their presence can be inferred from the women whirling around in their brightly colored dresses. Even more noticeably, the strong gravity of a black hole in such a system tears material from its companion star and funnels it into a disk around the hole's equator. Here, it gets compressed and heated, causing it to give off intense x-rays that can be detected by telescopes in Earth orbit.

Much larger black holes, weighing millions of solar masses, have also been discovered lurking in the centers of many galaxies. Astronomers have detected these supermassive black holes through measurements of bright stars close to the galaxies' centers. The stars are moving too fast and in orbits that are too tight for the central mass to be anything else but a black hole. Our Milky Way galaxy is thought to harbor a black hole in its core weighing around 2.6 million times the mass of the Sun. The biggest supermassive black hole, found in the heart of the distant galaxy Q0906+6930, weighs 10 billion times the mass of the Sun. This gives it an event horizon 60 billion kilometers across—200 times the size of Earth's orbit about the Sun.

Some bizarre physical effects take place around a black hole. For example, as an astronaut approaches the event horizon, time appears to slow down. An astronomer watching from a distance through a powerful telescope would see the astronaut's wristwatch tick slower—the effect is similar to the time dilation caused by traveling close to the speed of light, which we encountered in Chapter 3. On the horizon itself, this effect becomes infinite and time there appears to freeze. The distant observer would see the astronaut get closer and closer to the horizon but never actually cross it. The same effect stretches out the wavelength of the light from the astronaut, eventually shifting it outside of the visual spectrum so that she gradually fades from view.

From the astronaut's point of view, the process is much less serene. Plunging toward the event horizon, the traveler passes through what's known as the *photon sphere.* With a radius 1.5 times that of the event horizon, gravity here is strong enough to hold photons (particles of light) in circular orbits around the hole: Looking left or right, an astronaut at this distance from a black hole sees light that has traveled in a perfect circle around the hole—enabling her to see the back of her own head. Nearing the horizon, the gradient of the gravitational field starts to become considerable. Our unfortunate astronaut starts to notice the difference in the force between her head and her toes getting stronger and stronger.

As the astronaut dives into the singularity, things get ugly. The gravitational gradient becomes so severe that her head and feet are stretched far apart as she finally meets her fate in the singularity—a rather gruesome process that physicists have dubbed "spaghettification." This happens because the gravitational force increases the closer the astronaut gets to the black hole. In the modest gravitational fields present around the Sun and the planets of our Solar System, the effect is only noticeable over great distances. But an astronaut falling feet first into the intense gravity of a black hole will experience such a huge difference in force between her feet and her head that her body will be stretched out long and thin like spaghetti.

And so a black hole seems the last place in the Universe that you might want to set up home. Yet in "The Three Doctors," the renegade Time Lord engineer Omega is discovered living inside a black hole that he created in a supernova explosion. It's unlikely that you could ever live in the sort of black hole created in a supernova. But, amazingly, some physicists have suggested that it might (sort of) be possible to live inside a black hole that was much bigger—the size of our entire Universe.

Rather paradoxically, the bigger a black hole is, the gentler the spaghettification it causes will be. Averaged out, the density of a supermassive black hole at the center of a galaxy is less than the density of water. The combination of low density and large size means that an infalling astronaut doesn't experience extreme forces until she's well over the event horizon and deep inside it. Even though the astronaut is still doomed to hit the singularity, she can at least enjoy a (very) small stay of execution. Whereas the time from passing through the event horizon of a solar-mass black hole to hitting the singularity is a few ten-thousandths of a second, in the case of a galactic black hole this is lengthened to a second or so.

True, that's still not much help, but what if you could make a truly monstrous black hole, so big that you could live inside it without any worry of hitting the singularity within your lifetime—or even longer?

It turns out that you can. As we saw in the last chapter, our Universe is currently expanding, with its ultimate fate determined by its density. In particular, if the Universe is above a certain critical density it will eventually recollapse back down to a point, a singularity known as the Big Crunch, doing so in several tens of billions of years' time. If we did live in a supercritical density universe, then anything living for tens of billions of years would be destined to intercept the singularity in the future. There would be no escape. The Universe would, to all intents and purposes, be like the inside of a black hole.

"If the Universe was very dense it would eventually recollapse and all of what we see could end up hitting a singularity. This would be like what happens in a black hole," agrees Malcolm MacCallum, a relativity researcher at Queen Mary, University of London.

However, he adds that although this is technically correct, treating the Universe as a black hole isn't taken especially seriously by physicists. "It's not really meaningful in terms of the whole Universe. To be able to talk about a black hole one needs an outside—the Universe as a whole by definition has no outside."

The latest measurements, taken using NASA's WMAP spacecraft, seem to be consistent with the Universe having a density just above the critical value needed to make it recollapse in this way.

Power House

In the Fourth Doctor adventure "The Deadly Assassin," we discover that the Time Lords derive their power from a black hole called the Eye of Harmony, located on the Time Lord home world of Gallifrey. Black holes are an unlikely, but not impossible, power source. In 1969, mathematician Roger Penrose of the University of Oxford discovered how a futuristic civilization could harness one.

Penrose's method required a black hole that was rotating. These had already been investigated on paper in 1963 by the New Zealand–born mathematician Roy Kerr, who had found that as well as an event horizon, a rotating black hole also has an extra flattened spheroidal surface around it. Inside this zone, which Kerr named the "ergosphere," space itself would be swept around with the black hole as it rotated (see figure 15).

Penrose realized that because the ergosphere lies outside of the event horizon, it's possible to enter, get "spun up" by the rotation and then leave again—rather like grabbing hold of a spinning merry-go-round and then jumping off at higher speed. He imagined a future civilization taking advantage of this. However, his calculations showed that it's not quite as simple as hopping into the ergosphere and then emerging at a faster speed. He found that a spacecraft wanting to gain a boost from the rotation would need to shed some mass into the black hole first. This would create a kind of "recoil," rather like the way a gun kicks back toward you when you fire it, slowing down the spin of the black hole but speeding up the spacecraft.

Penrose suggested that this is one way in which an advanced alien race might dispose of its garbage. Shuttle vessels could carry the garbage into the ergosphere and unload it. He showed that in order to get a boost, they would have to unload the garbage onto a trajectory heading against the black hole's spin and in toward the event horizon. The shuttle would then receive a kick, accelerating it so that it emerged from the ergosphere with more energy than it, and its cargo of garbage, originally went

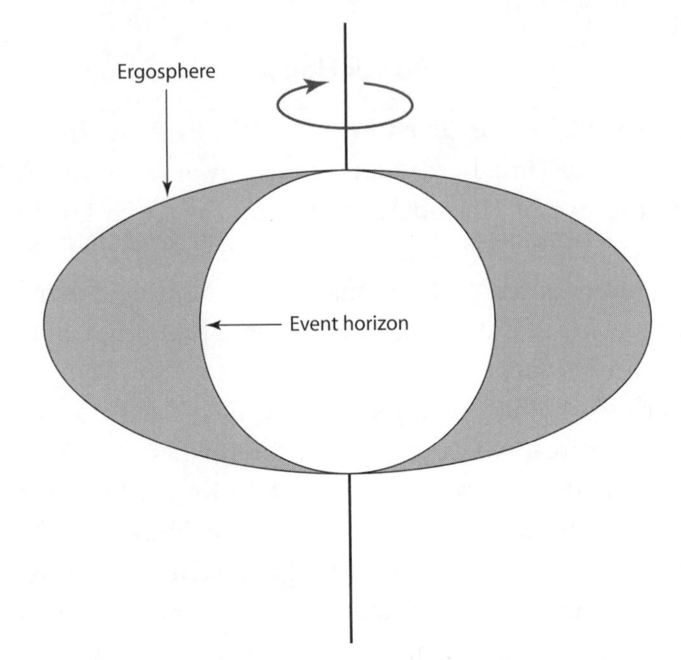

FIGURE 15. The ergosphere is a flattened spheroidal region surrounding the event horizon of a rotating black hole. Within the ergosphere, space is dragged around by the spinning hole. A wily future civilization could harness the energy of this rotation.

in with. And in return, the total energy of the black hole would decrease by a small amount.

The fast-moving empty shuttles would then need to dock with some kind of generator-like mechanism to convert their extra kinetic energy into electrical power. "In principle one could extract up to 29 percent of the mass-energy of the black hole in this way," says MacCallum. Using this technique, a single star-mass rotating black hole could provide a civilization with clean energy for thousands of years. "Although it would have to be a very advanced civilization to do this!" he adds.

He suggests that an alternative, and maybe slightly easier, way for the Eye of Harmony to work would be for the Time Lords to harness the energy given off when gas and dust falls into any black hole. This material—typically dragged in from a binary companion star or, in the case of galactic black holes, consisting of stars pulled in from the surrounding galaxy—forms a disk around the black hole and then spirals in, giving off ferocious amounts of energy in the form of jets squirted out at right angles to the disk. It's a process that's observed frequently by astronomers. "This is the standard explanation now for the jets seen in radio galaxies [galaxies that emit a significant amount of radio waves]," says MacCallum. "Certainly in science fiction one could imagine harnessing that energy. The disk would have a hole in the center and so would even look a bit like an eye."

Manipulation of black holes was taken to a whole new level by the Nimon race in the Fourth Doctor adventure "The Horns of Nimon," when they tried to create a black hole of their own. According to a report that appeared in *New Scientist* magazine in March 2005, scientists at the Relativistic Heavy Ion Collider (RHIC)—a huge particle accelerator at Brookhaven National Laboratory in Long Island, New York—now believe they are doing much the same thing.

Scientists are using the RHIC to collide beams of gold nuclei into each other at close to the speed of light. The terrific energy of the collisions breaks down the nuclei into a fireball of quarks (the smaller subatomic particles from which protons and neutrons are made up) and gluons (the particles that mediate what is called the strong nuclear force, the force of nature that binds quarks together). The fireball absorbs some of the other particles created in the collision. However, scientists find that ten times as many particles are being absorbed as calculations say there should be.

The researchers think that a tiny black hole is being created in the core of the fireball and that this is what's eating up the extra particles. They've compared the behavior of the fireball seen in their experiments with calculations of what they should see were a black hole to be present, finding good agreement.

Some scientists have worried that black holes created in particle accelerators like the RHIC could burrow down to the core of the Earth, from where they would proceed to devour the planet. But the black holes created in experiments such as the RHIC are tiny—lacking the gravity to swallow anything that's not fired straight at them. "I don't think there's any real danger," says MacCallum.

That was probably little consolation to the Doctor's colleague, the Brigadier, when UNIT HQ (with the Second Doctor, the Brigadier, and Sergeant Benton inside it) was devoured by a black hole in "The Three Doctors" and whisked away to an alternative universe.

BRIGADIER: "Now see here Doctor, you have finally gone too far!"

DOCTOR: "I rather think we all have. What's it like out there?"

BRIGADIER: "There's . . . well there's sand everywhere!"

DOCTOR: "Oh, splendid! Who's for a swim?"

BRIGADIER: "Do you realize what you've done? You've stolen the whole of UNIT HQ. Now what am I going to tell Geneva? That the whole blessed building has been picked up and put down on some deserted beach? We're probably miles from London."

DOCTOR: "I'm afraid we're a little bit further than that, Brigadier."

BRIGADIER: "You mean we're not even in the same country? There'll be international repercussions. This could be construed as an invasion."

BENTON: "It's not just a matter of the same country, sir. If the Doctor's right, we're not even in the same universe."

BRIGADIER: "What? Oh nonsense, Benton. That's a beach out there. It's probably Norfolk or somewhere like that."

DOCTOR: "Oh, please, if you'd only listen to me . . ."

32

Journeys through E-Space

"E-Space is another universe; there isn't a taxi service that goes back and forth."

—The Fifth Doctor, "Earthshock"

In October 1980, the Tardis went somewhere it had never been before. Dubbed E-Space, or the Exo-Space Time Continuum, by the *Doctor Who* writers, it's a "baby" universe appended to our own much larger universe, N-Space.

The Tardis was en route to Gallifrey, in the Fourth Doctor adventure "Full Circle." However, as it approached the yellowy planet it passed through a rare spacetime anomaly known as a *charged vacuum emboitment.* Similar in structure to a wormhole (see Chapter 3), this tunnel through space and time brought the Tardis to the planet Alzarius, a world with the exact same position in space as Gallifrey—except that its coordinates were negative. The Doctor stayed in E-Space for three adventures, returning to N-space in "Warrior's Gate," leaving his assistants Romana and K-9 behind but bringing new companion Adric back with him. Could there be a gateway to anything resembling E-Space appended to the universe that we live in?

Some astronomers and physicists think so. They're becoming increasingly struck by the idea that there's more to reality than the one Universe that we can see from Earth on a clear night. They believe that there's a sprawling network of hidden universes out there, and they've even got a name for them all: the "multiverse."

The term was first used in December 1960 by Andy Nimmo, then vice president of the Scottish branch of the British Interplanetary Society. Science fiction and fantasy author Michael Moorcock subsequently purloined the word for his Eternal Champion series of novels, and it has now made its way into standard scientific parlance.

The trouble is that scientists can't seem to make up their minds exactly what it means. A "collection of universes" is, after all, a rather broad definition. Astrophysicist Max Tegmark of MIT has come up with the best definition so far—or rather, a set of definitions. In an article published in the May 2003 issue of the magazine *Scientific American,* he set out four different "levels" of multiverse, all describing parallel universes existing beyond our own.

Parallel Worlds

Level I is a natural consequence of the theory of cosmic inflation, which we encountered in Chapter 30. As we saw, inflation is a phase of extremely rapid expansion in the very early history of our Universe, starting very soon after the Big Bang and finishing just 10^{-32} seconds later. Our Universe today is expanding gently, but back during inflation the expansion went into overdrive, increasing the size of the cosmos by an enormous factor of 10^{30}. Add to that 13.7 billion years (the time that has elapsed since the Big Bang) of conventional expansion and you start to see that our Universe is seriously huge.

Because of light-travel time, when we look out into space we see stars and galaxies as they were long ago. For example, we see the Andromeda galaxy, which is 2.2 million light-years away, as it was 2.2 million years ago. Similarly, when we look out into space the farthest point that we can see from Earth is 47 bil-

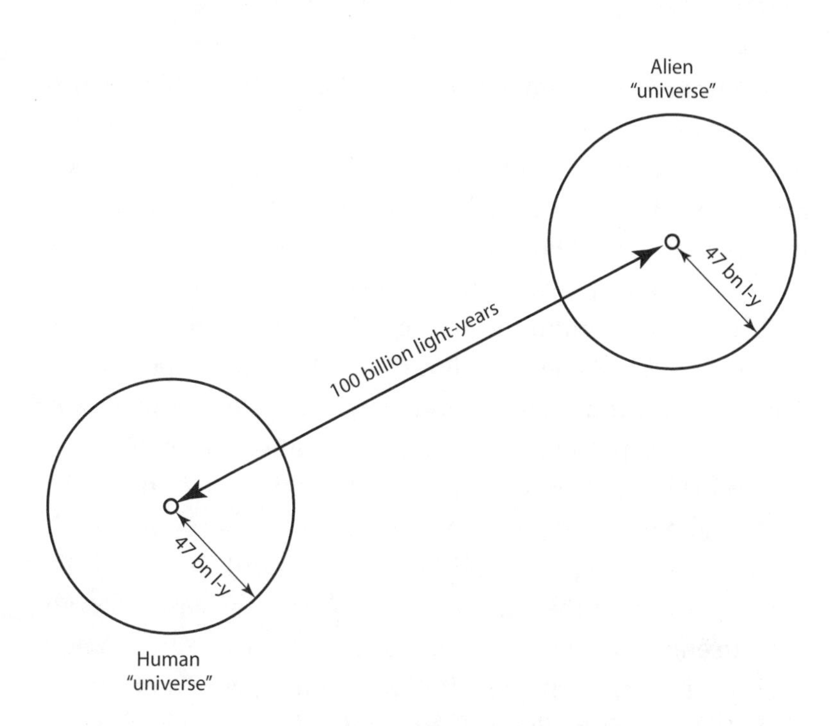

FIGURE 16. Because the universe began a finite time ago, there is a limit to how far we can see out into space—given by the distance that light can travel in this time, 47 billion light-years. And what we can see in our universe would be vastly different from what an alien species on a planet 100 billion light-years away would be able to see—for all intents and purposes, the two would be living in different universes altogether.

lion light-years away (see figure 16), with new regions of space rolling into view at the rate of 1 light-year per year. (In fact, we can't actually look this far because, as explained in Chapter 30, the Universe was opaque for its first 300,000 years, but let's assume for the sake of this thought experiment that we can.)

Now imagine an alien on a planet in a galaxy 100 billion light-years away. He, she, or it will be so far away that we on Earth won't have seen its galaxy, and it won't have seen our Milky Way. In fact, neither party is due to see the other's patch of space for another 53 billion years. The alien is, in effect, totally disconnected from us—as if it's living in another universe.

And this is the basic idea with a Level I multiverse. Because inflation has made space so big, the Universe is made up of a vast number of disconnected volumes, each of which can be thought of as an independent universe in its own right. Tegmark calculates that there are so many universes in a Level I multiverse that there's likely to be an exact copy of you, reading this very same book, on a planet just like Earth somewhere around 10 to the power of 10^{29} light-years away.

"There are an infinite number of other inhabited planets, including not just one but infinitely many that have people with the same appearance, name and memories as you, who play out every possible permutation of your life choices," he explains.

Level II is also based on inflation. But this time it stems from an idea about how inflation actually got going, proposed in the 1980s by cosmologist Andrei Linde of Stanford University. The problem was that the Universe was born in a hyperdense state, with an enormous gravitational field. How does it avoid collapsing under such terrific gravity and snuffing itself out of existence? The idea was that inflation saves the day—quickly blasting space up to a huge size before it has a chance to collapse. But how does the process get started? Researchers knew what was needed—the Universe had to be filled with "scalar fields" that had become trapped in a high-energy state (see Chapter 30). But how do you fill the entire early Universe with this kind of material?

Linde realized that you don't have to. He came up with an idea called *chaotic inflation,* because it supposed that the Universe started out in a chaotic, disordered state. Here, quantum fluctuations affected not just subatomic particles but also the very structure of space and time itself, making the Universe a seething cauldron of activity on the smallest scales. From point to point, fields and particles were taking wildly different values. And so from a purely statistical point of view, it stood to reason that there would be somewhere where the conditions were just right for inflation to start. And when it did start, this tiny region wasn't going to stay tiny for long—it would swiftly grow in size to become the dominant portion of the Universe.

Linde's theory of chaotic inflation led him naturally to another

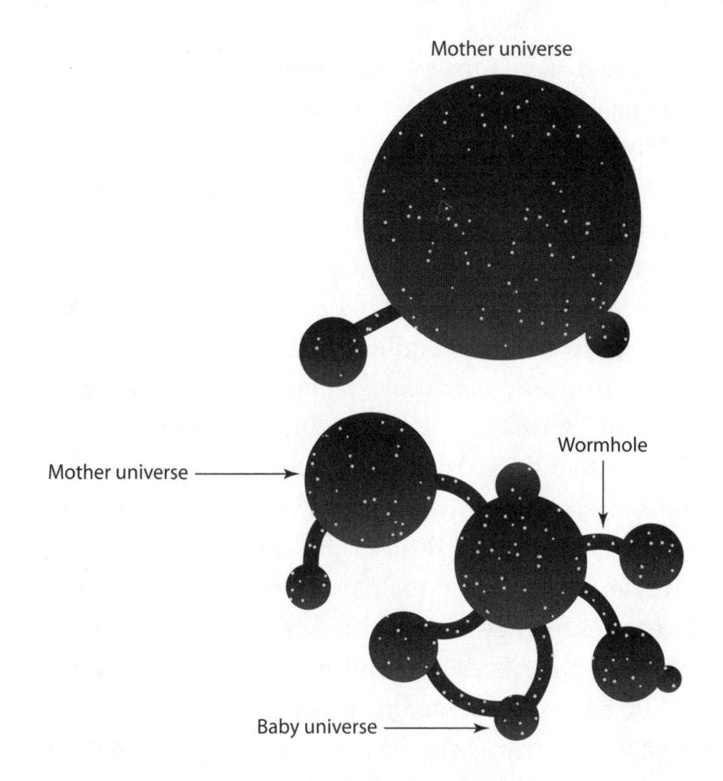

FIGURE 17. Physicist Andrei Linde has suggested that cosmic inflation is still happening today, causing small regions of space to bud off from our own Universe to form a network of interlinked baby universes.

even more mind-blowing idea, which he called *eternal inflation*. This says that chaotic inflation is still happening today—it's an ongoing process. Random quantum fluctuations in our Universe are constantly creating new inflationary universes, which are budding off from our own. And so a multiverse picture starts to emerge of an interlinked network of universes and baby universes (see figure 17). If this is correct, then our own Universe may actually be a baby universe—a bubble of spacetime that sprouted away from its parent universe 13.7 billion years ago. Interestingly, cosmologists think that the constants of nature—numbers in physics that determine things like the strength of gravitational and electromagnetic forces—might vary between universes in a Level II multiverse and that even the number of spatial dimensions could be different. This could lead to some bizarre forms of life (assuming that the constants and dimensions didn't vary so wildly as to make life altogether impossible).

If we really do live in a Level I or Level II multiverse, then it's possible that wormholes exist that link its different universes together. Some researchers have suggested that baby universes are created wherever black holes form in our Universe. And so if there was a way to survive falling through a black hole—maybe by turning it into a wormhole—then that could be a way for a hypothetical space traveler to ply the multiverse by jumping from bubble to bubble. If so, then E-Space—which is described in the show as a baby universe much smaller than our own, and so seems to fit the Level II description rather well—could be a very prescient creation on the part of the *Doctor Who* writers.

What about the Level III multiverse? This is exactly the kind of multiverse that Andy Nimmo was giving his talk about in 1960 when he coined the name. It comes from an idea in quantum physics known as the Many Worlds Interpretation, first proposed in 1957 by researcher Hugh Everett of Princeton University. The Many Worlds Interpretation says that an infinite number of parallel universes exists in which every conceivable version of reality is played out. What's more, new universes are being constantly created. So for example, when you toss a coin there will be some universes where the coin comes up heads, and some universes where it comes up tails, and there will be some universes where you tossed an eggplant in the air instead.

"Everything in our Universe—including you and me, every atom and every galaxy—has counterparts in these other universes," explains David Deutsch of the University of Oxford. "Some counterparts are in the same places as they are in our Universe, while others are in different places. Some have different shapes, or are arranged in different ways; some are so different that they are not worth calling counterparts."

That means that somewhere there are parallel universes in which you wrote this book, or where you are of the opposite sex, or where you don't even exist. Importantly, parallel universes in this view of the multiverse can "interfere" with our own Universe and others, and it's thought that this interference explains some of the key features of quantum physics, such as how particles can behave as waves and vice versa.

Whereas universes in Levels I and II occupy the same piece of space (but are just very far apart in it), in Level III each universe now represents a completely separate quantum reality. Could we ever jump from one such multiverse world to another? In a sense we already do. All the possible universes with a version of you in them occupy a broad swath in the ever-branching, ever-expanding multiverse. Whenever you do something, all of the possible outcomes actually happen—but each happens in a different universe. So if we look for a moment at one version of you, in one universe, who is just about to open their curtains in the morning, what do you see? The moment you open the curtains, the multiverse branches so that in one universe you might see a cat, in another you might see a bird, in another still you might see the postman, and so on. Which of these do you actually see? You see all of them. Before you opened the curtains you were in just one universe—afterward you are in many (one for each possibility), so in a sense you are constantly traveling into different universes.

Can we jump from one of these parallel universes to others via a wormhole and so visit our alternative selves? Deutsch thinks it might be possible. "If big enough wormholes existed, then, yes, some of them would lead to other universes, containing other copies of ourselves," he says.

Hyperspace Bypass

The final sort of multiverse—Level IV—is what Tegmark calls the "ultimate ensemble theory," because it allows not just for the physical constants and the dimensionality of alternate universes to vary (as in Level II) but also for them to be governed by completely different laws of physics. According to Level IV, every self-consistent mathematical theory for the Universe that can be formulated corresponds to a real universe somewhere.

If it sounds crazy, then—according to some physicists and mathematicians—that's because it is. "Tegmark seems to be saying that anything you can think of must exist, just because you can think of it," says Ian Stewart, a mathematician at the University of Warwick in England. "He's confusing the mathematical space of all the things that could happen with the physical space of all the things that actually do happen. If nothing else, this falls outside the normal range of what we consider to be science, because there's no way you can test it."

So it seems like it would be an unlikely event that could enable you to travel between two universes in a Level IV multiverse. And even if you could do it, you might not really want to. Doing this could land you in a universe where the laws and forces of nature are totally different to the ones that you're used to. Perhaps you would end up in a universe where electromagnetism works only over large distances—causing all your molecules to drift apart. Or maybe there would be a funny fifth force at work that causes electrons and protons to repel each other on small scales, disrupting the processes essential for life. With this much disparity possible, it's unlikely that we could ever hope to travel between two Level IV universes—or to survive the journey, at any rate. So we shan't dwell on Level IV here. Suffice it to say that as far as multiverses go, this is the daddy. And as all other multiverses must by definition be subsets of Level IV, there cannot be a Level V.

However it's done, venturing into uncharted parallel worlds is always going to be at best uncertain; at worst outright perilous. Indeed, the Fourth Doctor explored E-Space in the early 1980s trilogy "Full Circle," "State of Decay," and "Warrior's Gate"—and he's never been back.

33

Strange Stars and Mirror Planets

"E equals mC cubed."
"Squared."
"What?"
"E equals mC squared, not cubed."
"Not in the extra-temporal physics of the time vortex."

—Dr. Percival and the Master, "The Time Monster"

Every now and again, astronomers catch a glimpse of something through their telescopes that turns our cherished picture of the Universe on its head—be it the cosmic microwave background radiation (the echo from the Big Bang in which our Universe was created) or new and unusual planets orbiting faraway stars. Theoretical astrophysicists are no strangers to rocking the scientific boat, either—often throwing up bizarre ideas and scenarios that call our long-held beliefs about Life, the Universe, and Everything into question.

But scientists aren't the only ones. From quark stars to mirror planets to blobs of strange matter destroying the Earth, the *Doctor Who* writers have also done their share of dabbling in speculative research at the cutting edge of astrophysics and cosmology.

The Tenth Planet

In the Cybermen's debut adventure "The Tenth Planet," the Cyberman world of Mondas arrives in our Solar System from deep space. Mondas, we're told, was once the Earth's twin planet, but its long elliptical orbit carried it way out of the Solar System. Its inhabitants grew weak as their planet left the warmth of the Sun behind, and they began to add cybernetic modifications to their bodies to make them strong again. So it was that the Cybermen's race was born. Once they were done modifying themselves, they turned the wrenches on their planet—adding a propulsion system so that they could pilot it through space and back toward the Sun. "The Tenth Planet" tells the tale of what happened when they arrived back home.

Most of us learned in school that our Solar System has nine planets. So that would mean Mondas is number ten. But not any more. In January 2005, a team of astronomers led by Michael Brown of the California Institute of Technology announced the discovery of a new object known as 2003 UB313, which would later be given the more manageable name of Eris, orbiting at twice the distance from the Sun as the former outermost world, Pluto.

Other large ice worlds had already been detected in the outer Solar System, such as Sedna (also discovered by Brown and announced in November 2003). But these new discoveries were all smaller than Pluto, giving astronomers a fair excuse not to call them planets. Not so with Eris. Observations with NASA's Spitzer Space Telescope indicate that Eris is about 2,700 kilometers across—20 percent bigger than Pluto. The only reason it wasn't discovered sooner is because it's so much farther away.

So you might think that makes Eris the tenth planet, meaning in turn that Mondas would be the eleventh. Wrong again.

The discovery of Eris led astronomers to seriously reconsider what they really mean by the word "planet." One thing's for sure: if Pluto was discovered today, it wouldn't be one. Both it and Eris are large members of what astronomers call the Edgeworth-Kuiper belt—a disk of icy objects left over from the formation of the Solar System, circling the Sun out past the orbit of Neptune.

For this reason, astronomers at the International Astronomical Union's 2006 General Assembly in Prague voted to reclassify Eris and Pluto as "dwarf planets," along with two other Edgeworth-Kuiper belt objects—Haumea and Makemake—and Ceres, the largest of the asteroids. And so the upshot is that that there are now eight planets in the Solar System—and if a new planet were to be discovered today, it would be the ninth.

What else do we know about Mondas in Doctor Who? We're also told that it's the exact mirror image of the Earth. Particle physicists have found evidence for something they call "mirror matter." They say that matter that's a mirror image of the "normal" matter around us could be a natural consequence of what physicists call "symmetries." These are things that you can do to a subatomic particle that leave the laws governing its behavior unchanged. Common examples include rotational symmetry (particles spinning clockwise are governed by the same laws as particles spinning counterclockwise), and translational symmetry (move a particle from A to B and the physics underpinning its behavior remains the same).

But a problem emerged when physicists considered how particles behave under reflection—changing the particle into its mirror image. Instead of finding another symmetry, they found that this leads to a new particle that behaves very differently indeed. In 1956, physicists Tsung-Dao Lee at Columbia University and Chen Ning Yang of Princeton University's Institute for Advanced Study realized that reflectional symmetry could be restored to the particle world by introducing a new kind of matter—which they called mirror matter.

"Mirror matter is a kind of mirror image of normal matter," says Ray Volkas, a physicist at the University of Melbourne. "The internal physical laws [the "microphysics"] of mirror particles are a perfect copy of those for ordinary matter, except that the sense of left- and right-handedness is interchanged."

But the microphysics is all that's mirrored. There's no reason at all to expect that planets made of mirror matter will be mirror images of the planets we see in our Solar System, as was the case with the Cyberman world of Mondas—that is, that we should find mirror planets with backward continents and a backward United States that has New York City on the West Coast.

Even if Mondas could exist, mirror matter can—broadly speaking—interact with the rest of the matter in our Universe only through the force of gravity. (I say "broadly speaking" because there are certain circumstances under which normal matter and mirror matter can interact non-gravitationally; however, these are generally rare and the interactions are weak.) So if you were in a spaceship flying past a mirror planet, you would feel its gravitational attraction pulling you toward it, but that's about all. Its intermolecular forces wouldn't affect the intermolecular forces of the normal matter that your ship is made of, and so if you tried to land on the planet's surface you would simply pass straight through. You wouldn't even be able to see the planet because mirror matter interacts only with mirror light rays—not the normal light rays that your eyes are capable of detecting. (That said, normal-matter objects swept up by the mirror planet's gravity—such as meteoroids and particles from the solar wind that streams out from the Sun—would collect in a blob at the planet's center, which might be visible if you were close enough or if you had a high-power telescope.)

Even though exact copies of the Earth are unlikely, Paolo Ciarcelluti, a physicist at the University of L'Aquila in Italy, thinks that mirror worlds of the same type as the Earth are possible. Because the same microphysics operates on a mirror world (albeit between mirror-matter particles rather than particles of normal matter), he says it's quite feasible for such worlds to have breathable atmospheres, green grass, and blue seas. "It is possible to form atoms and molecules exactly as in our observable sector," he says. "And so even life could be possible, although this life would manifest itself in a form different from human life, as a result of the peculiar astronomical conditions and local evolution." Volkas agrees that Earth-like mirror worlds (as opposed to exact mirror images of Earth) are quite feasible.

Could mirror planets like this really exist wandering in our Solar System? Both Volkas and Ciarcelluti rule out the possibility of large mirror worlds orbiting our Sun, because by now we would have detected their gravitational influence on the other planets. "But small comet-sized or meteor-sized mirror matter objects could well exist in the Solar System," says Volkas. "One would also expect some amount of mirror gas."

For this reason, some scientists have speculated that mirror matter could make up some or all of the Universe's "dark matter." Studies of the gravitational motions of galaxies and clusters of galaxies show that there's vastly more matter in the Universe than can actually be seen. Bright, visible material such as stars accounts for as little as 5 percent of the Universe's total mass. The rest is made of unseen material, known as dark matter and dark energy (see Chapter 30 for more). No one knows for sure what either of these is actually made of, but mirror matter is one candidate for the dark matter content of the cosmos.

Although there's no solid evidence yet for the existence of mirror matter, no astronomical observations have ruled it out either. Researchers say that if it's out there then the tell-tale signs should be written into the structure of the Universe on every scale. "Mirror matter provides specific signatures on the formation of large-scale structures in the Universe, galactic structures, the cosmic microwave background radiation, and microlensing events," says Ciarcelluti. Microlensing events occur when the brightness of a background star is increased by the passage of an invisible mass in front of it (see figure 18).

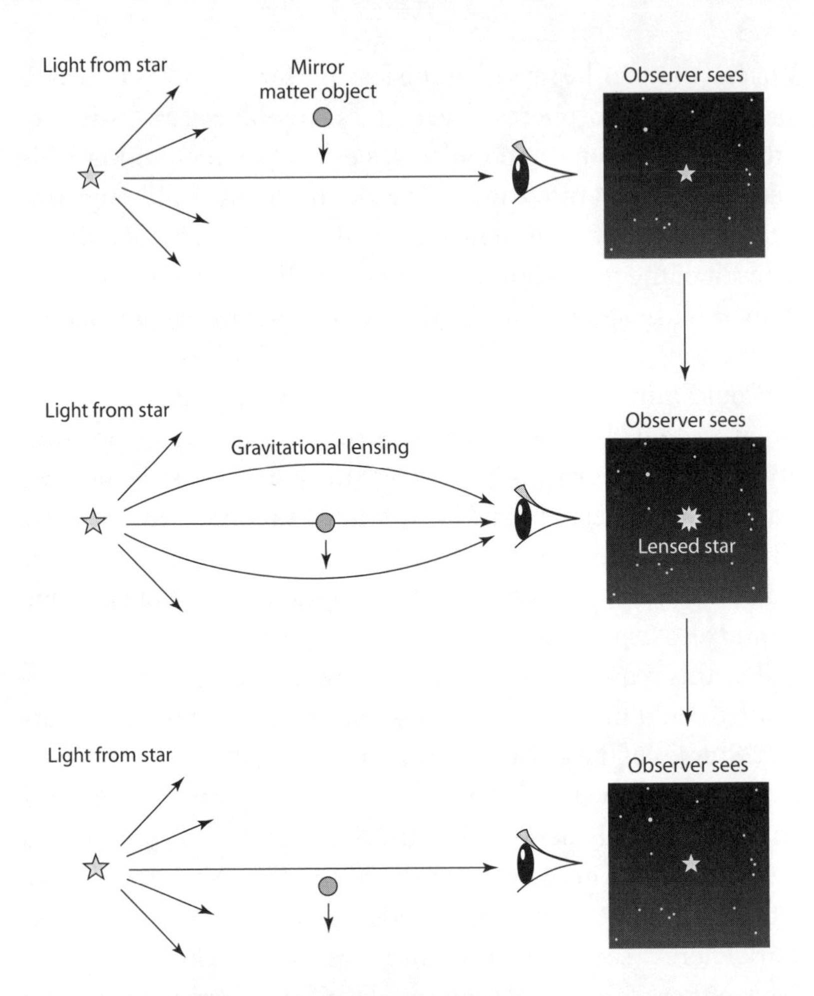

FIGURE 18. One way to detect otherwise-invisible mirror matter objects in our galaxy is through what is called gravitational lensing or microlensing. When a mirror matter star or planet moves across the line of sight between a normal star and an observer on Earth, its gravity bends the normal star's light, like a lens, causing the star to temporarily brighten.

Studies by astronomers of microlensing by objects in our own galaxy are consistent with as much as 50 percent of the galaxy's dark matter being made of these invisible microlensing objects—and if these really are made of mirror matter, then the galaxy is brimming over with the stuff.

Strange Brew

Mirror matter isn't the only bizarre hypothetical material that's had a mention in *Doctor Who*. In the Fifth Doctor story "Arc of Infinity," the Doctor encounters Rondel, which we are told is a collapsed "Q-star" and a strong source of magnetic radiation.

In fact, Q-stars, or quark stars to use their full name, are theoretical objects well known to astrophysicists (figure 19). They're thought to be the remnant state left behind after a very massive star has ended its life in a supernova explosion. In Chapter 31, we saw how a supernova can sometimes leave behind a neutron star, which is essentially the highly condensed core of the star that exploded. It's made of neutron particles packed together very tightly. But neutrons aren't the most fundamental kind of matter. Each neutron is itself made up of three smaller particles, known as quarks. And if the pressure exerted on a neutron star as it forms is great enough, then it's thought that the structure of individual neutrons nestled up against one another can actually melt away, so that either the whole star—or a spherical inner portion of it—could consist of one big blob of quarks.

If quark stars do exist, then they must tread a fine line—being denser than a neutron star, but not so dense that the whole thing collapses down unchecked to become a black hole (see Chapter 31). In 2002, a team of astronomers led by Jeremy Drake at the Harvard-Smithsonian Center for Astrophysics in Cambridge, Massachusetts, announced that they had found a strong candidate for a quark star using the Earth-orbiting Chandra X-Ray Observatory. The object, known as RXJ1856, was found to be just 11 kilometers across—far too small to be a neutron star, yet too bright to be a black hole. Drake cautions, however, that it's possible what his team are seeing is just a small, bright hot spot on the surface of a more usual neutron star.

It's also thought that quark stars will be highly magnetized, just like Rondel in "Arc of Infinity." You can think of a star's magnetic field as a bunch of lines threading through it; the closer together the lines, the stronger the field. When the star's core

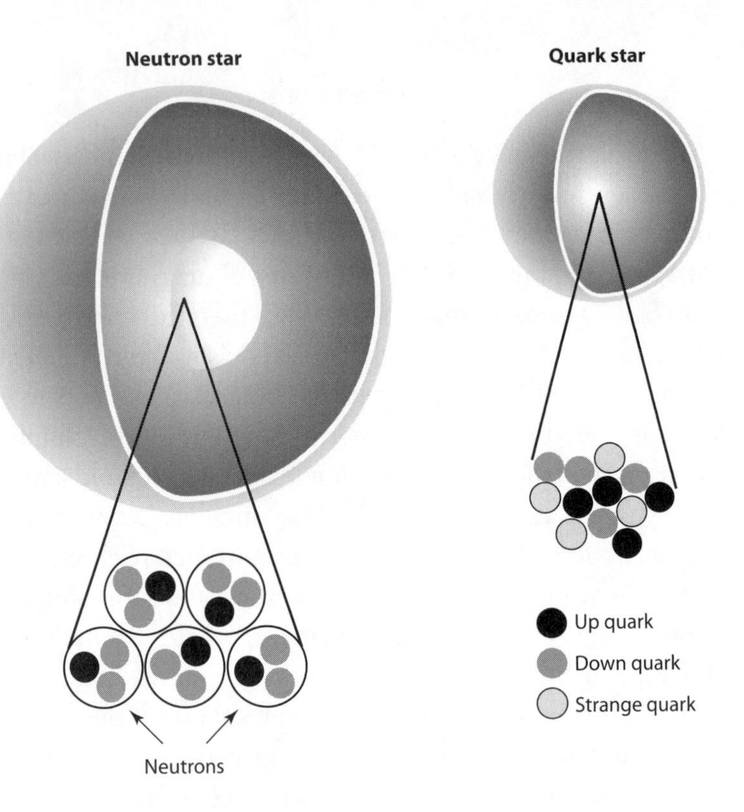

FIGURE 19. Quark stars and neutron stars are both possible remnant states left behind after a high-mass star has ended its life in a supernova explosion. A neutron star (*left*) is formed when extreme pressures in the core of the exploding star force protons and electrons together to leave a blob of neutrons about 25 kilometers across. But if the pressure is slightly higher, then even these neutrons melt into their constituent quarks to make a so-called quark star (*right*)—a ball of quarks just 10 to 20 kilometers across. Quarks come in three "flavors"—up, down, and strange—shown in the figure.

collapses, the magnetic field threading through it is condensed down, bunching the field lines together extremely tightly and sending the field strength through the roof. A typical massive star has a magnetic field about 100 times the strength of the field at the surface of the Earth. That's already fairly strong. But when this is packed down from stellar size to a sphere just 10 to 20 kilometers across—the size of a quark star—the field gets amplified by a terrific factor of some 10 billion, to create a field a trillion times stronger than the Earth's.

Physicists sometimes refer to the quark soup that quark stars are made from as "strange matter." This stuff has also enjoyed a cameo in *Doctor Who*. In the Seventh Doctor serial "Time and the Rani," the Rani (one of the Doctor's evil Time Lord foes) tried to blow up an asteroid made of strange matter, claiming that the resulting blast would destroy most of this corner of our galaxy.

Some physicists believe that it may be possible for asteroid-sized chunks of strange matter to exist outside of the phenomenal pressures inside a quark star. They refer to these chunks as "strangelets." If these objects exist, then they could be a source of concern—not because they might blow up the galaxy, but because any strangelet coming near the Earth might convert the Earth into strange matter as well, crushing us all out of existence. The basic idea is that the strong gravity of the superdense strangelet would squash down anything coming near it, until it too had the same high density. In this way, a renegade strangelet could "eat" its way through any stars or planets that it came into contact with.

These fears came to a head in 1999, when physicists prepared to switch on the RHIC particle accelerator at Brookhaven National Laboratory on Long Island. Some worried that the machine—which smashes atomic nuclei into one another at close to lightspeed—might create strangelets, and possibly mini-black holes too, that would then proceed to devour the planet and all on it. Similar fears resurfaced ten years later when the Large Hadron Collider was to begin colliding protons under France and Switzerland in an attempt to answer questions about the existence of type of particle called a Higgs boson.

Back at the time that the RHIC was being switched on, Robert Jaffe of MIT reassured readers of *New Scientist* magazine: "Strangelets are almost certainly not stable, and if they are, they almost certainly cannot be produced at RHIC. And even if they were produced at RHIC, they almost certainly have positive charge and would be screened from further interactions by a surrounding cloud of electrons."

The RHIC was activated in 2000 and, well—so far, so good. In April 2005, the collider succeeded in producing a high-energy soup of quarks in an attempt to recreate the conditions of the early Universe. If any strangelets were produced, it seems, thankfully, that none of them escaped. And if they did, they don't seem to have eaten too much of the planet just yet. So, it's back to the drawing board, Rani.

34

The More Things Change

"Strictly speaking, it's the fifteenth New York, which makes it New New New New New New . . . New New New New New New . . . New New New York."
(Rose laughs)
"What?"
"You're so different."
"New New Doctor."

—The Tenth Doctor and Rose, "New Earth"

After a stunningly successful return to British TV in 2005, with Christopher Eccleston in the title role, *Doctor Who* returned once more in 2006, with accomplished Scottish actor David Tennant as the Tenth Doctor. And, just like many of the Doctors before him, the Tenth Doctor found himself battling the Cybermen.

The original Cybermen of the classic series were from the planet Mondas, a twin world to the Earth that had drifted off into space. As the errant world wandered farther from the life-giving energy of the Sun, its inhabitants turned to cybernetic implants to keep them alive. Eventually, the Mondassians became more machine than human.

But the metal menaces that David Tennant's Tenth Doctor first encounters in 2006's "Rise of the Cybermen" were a new breed altogether. They were constructed on a version of the Earth existing in a parallel universe. The man responsible was John Lumic, the head of Cybus Industries. Lumic was terminally ill and had developed the cyborg technology in order to keep his own ailing body alive. But when Britain's president (this is a parallel world, remember) learns of Lumic's plans, he is horrified. He refuses to let the research continue. Outraged, Lumic unleashes his creations. Their instructions: to take over the planet.

The major difference between Lumic's Cybermen and their Mondassian predecessors is that the Mondas Cybermen had fundamentally organic bodies that had been augmented with cybernetic plug-ins. Lumic's "ultimate upgrade," however, took just the brain from a human subject and then encased it in an entirely robotic metal body.

This is certainly feasible. True, attempts to perform brain transplants between living creatures, as described in Chapter 22, have invariably left their recipients quadriplegic—paralyzed from the neck down. This is because reconnecting fibers of the central nervous system is, at present, impossible. But if all the connections to the brain are made artificially, side-stepping the need to work with living nervous system tissue, then brain transplants suddenly become more plausible. As we saw in Chapter 9, controlling electronic equipment directly from the human brain has already been achieved.

Not only is this approach to making Cybermen easier, but the final result could also be superior. The new Cybermen have fewer fragile organic body parts, making them less vulnerable to attack. They also do not appear to have the weakness to the precious metal gold that so afflicted the Mondassian Cybermen in earlier *Doctor Who* adventures. Maybe this Achilles heel was a consequence of the Mondas Cybermen having respiratory systems that are partially organic—something Lumic avoided with his totally robotic body design.

Although these nouveaux Cybermen are impervious to gold, the Doctor still manages to find and exploit a weakness in their one remaining organic body part: the brain. Waking up to find that someone has replaced your flesh and blood with a steel exoskeleton is enough to tip even the most rational souls over the edge. So to prevent his Cybermen from going insane at their predicament, Lumic fitted them with "emotional inhibitor" chips, which literally block the brain's emotional responses.

Our emotions emerge from a region of the brain called the limbic system. It's buried deep within the brain, and in evolutionary terms is one of its oldest components. Maybe Lumic's emotional inhibitors could work by blocking the limbic system's activity in some way. Indeed, according to J. Reid Meloy, a forensic psychologist at the University of California, San Diego, the emotionless state of some psychopaths and serial killers may be the result of damage to or malformation of the limbic system.

But how might you go about doing this? Historically, brain surgeons have severed nerve connections between certain parts of the brains of mental patients in grisly procedures known as lobotomies. The technique was "pioneered" by Portuguese neurosurgeon Egas Moniz, who drilled holes in patients' heads and then injected alcohol to destroy certain regions of the tissue inside. Other, even more barbaric techniques involved the insertion of long thin blades into holes in the patient's skull, which were then used to physically sever the connecting brain tissue. There was little intention to cure patients by this method—simply to make them more docile and easier to handle by asylum staff. Mercifully, lobotomy operations are rarely performed today.

Lumic would not have wanted to risk damaging his subject's brains by cutting into them—especially since the seat of emotion, the limbic system, lies so deep within. But chances are he wouldn't have needed to. He has already mastered how to harness the brain's electrical output and use it to control his Cybermen's mechanical bodies. So it also seems likely that he would have figured out how to block any brain impulses that he wanted to suppress. Perhaps long, needle-like electrodes would have penetrated the limbic system. There, moderated by the emotional inhibitor chip, they would deliver flashes of electrical current to nullify emotional responses as and when they happened.

In the 2006 season climax "Doomsday," the Doctor turns the emotional inhibitor against the Cybermen. However, if some social psychologists are to be believed, this intervention might not have been needed. Antonio Damasio, of the University of Southern California, says it is the reaction to emotions that motivates virtually all of our actions and that without emotions we would be incapable of decision making. Another team of researchers, at the University of Melbourne, has found that blocking emotions can lead to depression. In 2006, they studied 50 people who had experienced at least one episode of depression in their lives. They found that those who tried to suppress unhappy memories and emotions were more likely to relapse into a depressive state than those who openly experienced them.

So it could be that emotions—sometimes seen to "fog" our decision-making capability—could be a crucial mental function for any species seeking cosmic domination. When Lumic deprived his Cybermen of their emotional responses, that could have been enough to ensure their downfall.

Of Wolf and Man

Werewolves—creatures that can shape-shift between human and wolf form—first appeared in the *Doctor Who* TV series in the 1988 Seventh Doctor serial "The Greatest Show in the Galaxy." One of the creatures was used again to great effect in Tenth Doctor adventure "Tooth and Claw," a gothic horror story that harks back to the earlier golden years of the show's Tom Baker era.

In "Tooth and Claw," the werewolf is explained as an alien life-form that fell to Earth in Scotland in 1540. By passing its cells from host to host—by biting victims—the "wolf" is able to survive for hundreds of years. Indeed, the serial is set in the year 1879, almost 350 years after the creature's arrival on Earth.

Werewolf legends are common in real-world mythology, and scientists and historians have come up with a number of theories to explain why this should be. One leading contender is rabies. Those infected with this disease become aggressive, foam at the mouth, and have an increased tendency to bite because of the spasmic response of the central nervous system that the virus causes.

Another disease that sometimes gets the blame for werewolf legends is porphyria. Sufferers of this illness can't produce the iron-rich heme molecules in their blood, resulting in increased sensitivity to sunlight, excessive hair growth, and disfiguration and even causing their gums to recede, making their teeth look longer. The same disease has also been invoked to explain vampire legends—some sufferers even exhibiting an aversion to garlic.

There's even a medical condition that goes by the pseudonym "werewolf syndrome." Hypertrichosis is a genetic condition that causes sufferers to grow extreme amounts of body hair, sometimes resulting in a thick coat of animal-like hair covering most of the face and body. Unfortunate people afflicted with this condition were sometimes recruited by circuses and freak shows during the early years of the twentieth century. However, although labeled with names such as "wolf boy," sufferers showed none of the aggressive symptoms that are usually associated with werewolves.

A more recent theory ascribes werewolf tales to cases of ergot poisoning. Ergot is a kind of fungus that grows on grains and grasses. When ingested it causes a number of symptoms, from restricted blood circulation (sometimes leading to loss of limbs) to seizures, hallucinations and insanity. The fungus could contaminate whole stores of grain, leading to mass incidents of ergot poisoning affecting entire towns. It's been suggested that the mind-altering affects of the condition could lead suffers to believe that they have seen a werewolf or to even become convinced that they are one themselves.

An even stranger notion still is the condition of "clinical lycanthropy." This is a rare psychiatric disorder, in which sufferers become delusional and believe that they possess the ability to change into animal form. According to a study by MacLean Hospital, a psychiatric unit affiliated with the Harvard Medical School, the condition is just one symptom of a more general psychosis in which patients experience delusions and hallucinations. Researchers Harvey Rostenstock and Kenneth R. Vincent, writing in the *American Journal of Psychiatry,* suggest that suffers of clinical lycanthropy could pose a serious threat to others and, in less enlightened times, would have been feared for their tendency to commit bestial acts and may even have been hunted down and killed.

These explanations are all rather less colorful than the possibility that werewolves were real creatures and that they came from outer space, as put forward in the "Tooth and Claw" episode of *Doctor Who*. But until we have convincing evidence for the existence of extraterrestrial life, these are the best theories that science has to offer.

Which is most likely? Choosing between these ideas is difficult because the evidence for each is limited, and it could be that this situation isn't going to change any time soon. "The problem is there's not a great deal of research in this area, because it would be frowned upon for anyone with a serious medical background to look into this in detail," Jayney Goddard, president of the UK-based Complementary Medical Association, said in an interview in the Autumn 2005 issue of *BBC Focus* magazine.

If that's the case then the truth about werewolf legends is a mystery that we may never get to the bottom of.

And speaking of things we will never really get to the bottom of—and I mean *never*—let's finish up with one of the subjects closest to the Doctor's heart: eternity.

35

The End of Time

"I'm a Time Lord . . . I'm not a human being; I walk in eternity."

—The Fourth Doctor, "Pyramids of Mars"

George Bernard Shaw once said that because the English aren't very spiritual people, they invented the game of cricket to give themselves some sense of eternity. Perhaps it also underpins our preoccupation with trains, the Internet, call centers—the list goes on for, well . . . eternity.

So what is it? Is it just a very, very, very long time? Or is there more to it than that? There's a commonly recited story that's meant to illustrate the concept of eternity. The Time Lord version of it goes something like this. Somewhere on the Doctor's home planet of Gallifrey (before the Daleks blew it up, obviously), there's a very tall mountain, taller than Mount Everest, and made of solid granite. Once every million years a tiny bird lands on the top of the mountain and touches it with a feather. Eternity, it's said, is the time it takes for the mountain to wear down to nothing. Isn't it?

"Well, actually, no," says mathematician Ian Stewart, of the University of Warwick, England. "That's still finite and completely misses the point. Eternity is not a very long finite time.

That's no closer to eternity than one year. Eternity is what happens when the mountain never wears down." The trouble is that if you pick a very, very large—but finite—length of time and call it eternity, then you can always add 1 second to it and get a period of time that's even longer. You know when you're dealing with the real eternity because that's when eternity plus 1 second is still just eternity. Adding or subtracting finite numbers makes no difference. Still with me?

Good, because eternity gets weirder than that. Much weirder. Imagine you have an infinite number of lottery balls. But rather than just being numbered 1 to 49, these balls are labeled 1, 2, 3, 4, and so on, forever. On day one, you take the balls labeled 1 to 10 and put them into a bag, but then you take out ball number 1 and throw it away. Similarly, on day two you take out ball number 2 and add balls 11 to 20. On day three, you take out ball 3 and add balls 21 to 30. Every day you take out one ball and add ten, so the number in the bag goes up by nine. In general, on the nth day, you take out ball number n and add ten balls of up to 10n in size.

After eternity has passed, how many balls are there in the bag? "The number is 9, then it's 18, then 27—it's going up all the time. But what happens after eternity is that it's empty," says Stewart. "If it's not empty then there must be a ball in it, let's say ball number 1,000,003—but you took that out on day 1,000,003, so it can't be there."

Confused? It gets better yet. If on day one you put in balls 1 to 10 and take out ball 10, then on day two you take out ball 20 and put in nine more, then on day three you take out ball 30 and put in more, and so on—then only balls that are a multiple of 10 get taken out, and so a lot get left in—this time, the bag definitely isn't empty at eternity.

Now comes the really clever bit. Suppose you repeat the procedure above except that the numbers on the balls are written in invisible ink. So you don't know which ones you're putting in and taking out—only that each day you put in ten new ones and take one out.

The end result seems to depend on exactly which balls you take out. Whether you end up with no balls in the bag, a finite number of balls in there, or an infinite number depends on which balls are taken out and the order they're drawn in. "But then if the numbers on them are written in invisible ink and all the balls look the same, you can't actually tell the difference," says Stewart. It's a bizarre paradox—and one to which he knows no resolution.

This is a great illustration of the power of eternity to bend our heads. But if mathematicians struggle to get to grips with the notion of eternity on a purely theoretical footing, then imagine what physicists have to go through when it crops up in calculations describing what's meant to be the real world. It certainly happens. As we saw in Chapter 31, a distant observer watching a hypothetical astronaut falling into a black hole sees the astronaut take an infinite amount of time to cross the event horizon. And as we saw in Chapter 30, if the density parameter of the Universe, Ω, is less than 1, then not only is the Universe infinite in space—it can last for an infinite time as well.

Some have even speculated that whatever went on before the Big Bang, in which the Universe was created, can be thought of as a kind of "eternity." The Big Bang created time as well as the matter and space that the Universe is made of, so that beforehand there was no time—at least not in the finite sense in which we normally think of it. Maybe what was there can be described in some sense as eternity.

Others, like me, have wondered whether their limited brains are really up to the job of comprehending such stupendous notions as infinity and eternity. "These are tricky concepts because they're idealizations of mental thought processes as much as anything," admits Ian Stewart.

Indeed, it's easy for us to be sloppy about what we mean by infinity and to use the word as a catch-all term even when we don't really know what we're talking about. Stewart cites the commonly uttered statement, "in an infinite universe everything possible would happen somewhere," as one of the worst misconceptions. “Then how about an infinite universe consisting of large numbers of copies of a chair?” he wonders.

Yet his prevailing view is one of optimism—that humans are eminently capable of getting a handle on the infinite and the eternal. Mathematicians with their careful formalisms and rigorous logical procedures are, perhaps like the Doctor, better equipped than most to understand it. As for the rest of us? Keep trying. After all, even if it does take a very, very, very long time to figure it out, it'll never quite take an eternity.

Epilogue

Tom Baker famously said in his final scene as the Fourth Doctor: "It's the end . . . but the moment has been prepared for."

As the deadline for this book looms and I somewhat hurriedly get my manuscript together, preparedness isn't exactly shining through as my most abundant quality. Who was it who said "organization is the enemy of creativity"? Actually, I think that was me—just then. Still, disordered thinking is an established Time Lord tradition observed by the Doctor on numerous occasions that frequently carried him along the path to true greatness. And if it works for him, then it's certainly good enough for me.

I hope you have enjoyed what you've read here. For me, the best news this decade is that *Doctor Who,* an institution that I grew up with, is back on our TV screens and looks as if it's here to stay. If this book can contribute to the phenomenon, and its continued success and longevity, in even the smallest way, I will be very happy indeed.

It's probably somewhere around here too that I'm meant to say something profound about the noble pursuit of science and tell you all to go away and enlist in correspondence courses in molecular biology to deepen your appreciation of the world around you. I'm not going to do that. This book was written first and foremost to entertain, to boost enjoyment of the show, and to answer questions that it may have raised in the minds of intelligent fans. I hope I've fulfilled those aims. If I did manage to educate anyone along the way, then I sincerely apologize.

Anyway, there's a timeline that needs preserving here. If I didn't finish it, you can't be reading it—so I'd best be getting on.

And whatever you do, keep watching.

Sorry, what did you say?

Yes I can. Look. Here I am. This is me swanning off.

See ya . . .

List of Episodes by Doctor

FIRST DOCTOR
"An Unearthly Child"
"The Daleks"
"The Edge of Destruction"
"Marco Polo"
"The Keys of Marinus"
"The Aztecs"
"The Sensorites"
"The Reign of Terror"
"Planet of Giants"
"The Dalek Invasion of Earth"
"The Rescue"
"The Romans"
"The Web Planet"
"The Crusade"
"The Space Museum"
"The Chase"
"The Time Meddler"
"Galaxy 4"
"Mission to the Unknown"
"The Myth Makers"
"The Daleks' Master Plan"
"The Massacre"
"The Ark"
"The Celestial Toymaker"
"The Gunfighters"
"The Savages"
"The War Machines"
"The Smugglers"
"The Tenth Planet"

SECOND DOCTOR
"The Power of the Daleks"
"The Highlanders"
"The Underwater Menace"
"The Moonbase"
"The Macra Terror"
"The Faceless Ones"
"The Evil of the Daleks"
"The Tomb of the Cybermen"
"The Abominable Snowmen"
"The Ice Warriors"
"The Enemy of the World"
"The Web of Fear"
"Fury from the Deep"
"The Wheel in Space"
"The Dominators"
"The Mind Robber"
"The Invasion"
"The Krotons"
"The Seeds of Death"
"The Space Pirates"
"The War Games"

THIRD DOCTOR
"Spearhead from Space"
"The Silurians"
"The Ambassadors of Death"
"Inferno"
"Terror of the Autons"
"The Mind of Evil"
"The Claws of Axos"
"Colony in Space"
"The Daemons"
"Day of the Daleks"

“The Curse of Peladon”
“The Sea Devils”
“The Mutants”
“The Time Monster”
“The Three Doctors”
“Carnival of Monsters”
“Frontier in Space”
“Planet of the Daleks”
“The Green Death”
“The Time Warrior”
“Invasion of the Dinosaurs”
“Death to the Daleks”
“The Monster of Peladon”
“Planet of the Spiders”

FOURTH DOCTOR

“Robot”
“The Ark in Space”
“The Sontaran Experiment”
“Genesis of the Daleks”
“Revenge of the Cybermen”
“Terror of the Zygons”
“Planet of Evil”
“Pyramids of Mars”
“The Android Invasion”
“The Brain of Morbius”
“The Seeds of Doom”
“The Masque of Mandragora”
“The Hand of Fear”
“The Deadly Assassin”
“The Face of Evil”
“The Robots of Death”
“The Talons of Weng-Chiang”
“Horror of Fang Rock”
“The Invisible Enemy”
“Image of the Fendahl”
“The Sun Makers”
“Underworld”
“The Invasion of Time”
“The Ribos Operation”
“The Pirate Planet”
“The Stones of Blood”
“The Androids of Tara”
“The Power of Kroll”
“The Armageddon Factor”
“Destiny of the Daleks”
“City of Death”
“The Creature from the Pit”
“Nightmare of Eden”
“The Horns of Nimon”
“Shada”
“The Leisure Hive”
“Meglos”
“Full Circle”
“State of Decay”
“Warriors’ Gate”
“The Keeper of Traken”
“Logopolis”

FIFTH DOCTOR

“Castrovalva”
“Four to Doomsday”
“Kinda”
“The Visitation”
“Black Orchid”
“Earthshock”
“Time-Flight”
“Arc of Infinity”
“Snakedance”
“Mawdryn Undead”
“Terminus”
“Enlightenment”
“The King’s Demons”
“The Five Doctors”
“Warriors of the Deep”
“The Awakening”
“Frontios:
“Resurrection of the Daleks”
“Planet of Fire”
“The Caves of Androzani”

SIXTH DOCTOR

“The Twin Dilemma”
“Attack of the Cybermen”
“Vengeance on Varos”
“The Mark of the Rani”
“The Two Doctors”
“Timelash”
“Revelation of the Daleks”
The Trial of a Time Lord: Part I. The Mysterious Planet (four-episode series)
The Trial of a Time Lord: Part II. Mindwarp (four-episode series)
The Trial of a Time Lord: Part III. Terror of the Vervoids (four-episode series)

The Trial of a Time Lord: Part IV. The Ultimate Foe (four-episode series)

SEVENTH DOCTOR

"Time and the Rani"
"Paradise Towers"
"Delta and the Bannermen"
"Dragonfire"
"Remembrance of the Daleks"
"The Happiness Patrol"
"Silver Nemesis"
"The Greatest Show in the Galaxy"
"Battlefield"
"Ghost Light"
"The Curse of Fenric"
"Survival"

EIGHTH DOCTOR

Doctor Who: The Movie

NINTH DOCTOR

"Rose"
"The End of the World"
"The Unquiet Dead"
"Aliens of London"
"World War Three"
"Dalek"
"The Long Game"
"Father's Day"
"The Empty Child"
"The Doctor Dances"
"Boom Town"
"Bad Wolf"
"The Parting of the Ways"

TENTH DOCTOR

"The Christmas Invasion"
"New Earth"
"Tooth and Claw"
"School Reunion"
"The Girl in the Fireplace"
"Rise of the Cybermen"
"The Age of Steel"
"The Idiot's Lantern"
"The Impossible Planet"
"The Satan Pit"
"Love & Monsters"
"Fear Her"
"Army of Ghosts"
"Doomsday"
"The Runaway Bride"
"Smith and Jones"
"The Shakespeare Code"
"Gridlock"
"Daleks in Manhattan"
"Evolution of the Daleks"
"The Lazarus Experiment"
"42"
"Human Nature"
"The Family of Blood"
"Blink"
"Utopia"
"The Sound of Drums"
"Last of the Time Lords"
"Voyage of the Damned"
"Partners in Crime"
"The Fires of Pompeii"
"Planet of the Ood"
"The Sontaran Strategem"
"The Poison Sky"
"The Doctor's Daughter"
"The Unicorn and the Wasp"
"Silence in the Library"
"Forest of the Dead"
"Midnight"
"Turn Left"
"The Stolen Earth"
"Journey's End"
"The Next Doctor"
"Planet of the Dead"
"The Waters of Mars"
"The End of Time"

Further Reading

Astrobiology: A Brief Introduction, by Kevin W. Plaxco and Michael Gross (Johns Hopkins University Press, 2006)

Bang! The Complete History of the Universe, by Brian May, Patrick Moore, and Chris Lintott (Johns Hopkins University Press, 2008)

Black Holes and Time Warps: Einstein's Outrageous Legacy, by Kip S. Thorne (Norton, 1995)

Black Holes, Wormholes & Time Machines, by Jim Al-Khalili (Taylor & Francis, 1999)

Dark Side of the Universe: Dark Matter, Dark Energy, and the Fate of the Cosmos, by Iain Nicolson (Johns Hopkins University Press, 2007)

Doctor Who: The Legend Continues, by Justin Richards (Random House UK, 2005)

The Fabric of Reality: The Science of Parallel Universes and Its Implications, by David Deutsch (Penguin, 1998)

The Fabric of the Cosmos—Space, Time, and the Texture of Reality, by Brian Greene (Vintage, 2005)

The Language of the Genes: Solving the Mysteries of Our Genetic Past, Present, and Future, by Steve Jones (Anchor, 1995)

The Luck Factor: The Four Essential Principles, by Richard Wiseman (Miramax, 2004)

Mars—The Inside Story of the Red Planet, by Heather Couper and Nigel Henbest (Headline, 2001)

Nanotechnology for Dummies, by Richard Booker and Eric Haroz (Wiley, 2005)

The Singularity Is Near—When Humans Transcend Biology, by Ray Kurzweil (Penguin, 2006)

Time of Our Lives—The Science of Human Aging, by Tom Kirkwood (Oxford University Press, 2002)

What Does a Martian Look Like: The Science of Extraterrestrial Life, by Jack Cohen and Ian Stewart (Wiley, 2005)

Index